Placer Gold Deposits of New Mexico

By MAUREEN G. JOHNSON

GEOLOGICAL SURVEY BULLETIN 1348

A catalog of location, geology, and production, with lists of annotated references pertaining to the placer districts.

CONTENTS

PLACER GOLD DEPOSITS OF NEW MEXICO

By Maureen G. Johnson

ABSTRACT

Thirty-three placer districts in New Mexico are estimated to have produced a minimum of 661,000 ounces of placer gold from 1828 to 1968. The location, areal extent, past production, mining history, and probable lode source of each district are summarized from a wide variety of published reports relating to placer deposits. An annotated bibliography of all reports that give information about individual deposits is given for each district.

Most placer gold deposits in New Mexico are derived from gold-bearing mineralized areas in Tertiary intrusive rocks, and occur in gravels of alluvial fans, gulches, and rivers adjacent to the source. A few deposits are derived from gold-bearing Precambrian crystalline or Tertiary volcanic rocks. Most of the major placer districts were discovered and extensively worked between 1828 and 1880; in later years, large-scale dredging operations were successful at a few localities, while intermittent activity continued at most districts.

INTRODUCTION

HISTORY OF PLACER MINING IN NEW MEXICO

Placer mining began in 1828, when the rich placer deposits of the Ortiz Mountains, Santa Fe County, were discovered (Jones, 1904, p. 21). Even before that discovery, New Mexico was the scene of some mining activity. The Pueblo Indians mined and used turquoise for ornaments, and there is some evidence to suggest that they used gold (probably collected from gravel deposits). Alvar Nunez Cabeza de Vaca, who stayed with the Pueblo Indians on his journey from Florida to Mexico, spread stories of gold and silver possessed by these Indians. Coronado left Mexico in 1540 with a large expedition to find and exploit the source of this supposed vast wealth. He found no great riches, and, after exploring the land, returned to Mexico in 1542. The Spanish returned to the territory in the late 1500's, established missions, and may have done some mining, for after the Pueblo revolt in 1680 the Indians stipulated that the Spaniards were not to engage in mining but were to confine their activities to agriculture. Placers were reportedly worked along the Rio Grande (Taos County) in the 1600's. The real

1

development of mineral resources began in 1800 after an Indian disclosed the location of the rich Santa Rita copper deposits to a Spanish officer. The first shipments of copper to Mexico were made in 1801.

Gold was discovered in the Ortiz Mountains at Old Placers in 1828. New Placers was discovered in 1839, and before 1846 minor deposits of placer gold were found at Taos and Abiquiu and in the Sangre de Cristo Mountains (Prince, 1883, p. 243). The placers along the Rio Grande were probably mined intermittently from 1600 to 1828—placer gold was found at Rio Hondo in 1826—but it was the discovery in the Ortiz Mountains that marked the beginning of real interest in New Mexico placers. During the decade 1860–70, many placer deposits were found and exploited, including the rich deposits at Elizabethtown (Colfax County) and Pinos Altos (Grant County). Many discoveries of rich lode deposits followed as a result of the placer discoveries. By the end of the 19th century, many of the placers discovered were already exhausted. In 1901 a prospector located the placers at the foot of the Caballos Mountains in Sierra County and tried (unsuccessfully) to keep the location of his rich find a secret. By 1903 the secret was out, and a gold rush to the Caballos Mountains followed. In 1908 the discovery of placers at Sylvanite (Hidalgo County) caused the last gold rush in New Mexico.

Development of some placer deposits, and abandonment of many others, continued until the depression years of the early 1930's. Placer mining all over the West underwent a great revival during the depression; many individuals turned to placer deposits to earn a grub stake or just a meal ticket. During this period much attention was given to the invention and development of a myriad of jigs, drywashing machines, and separation methods for recovery of gold from placers and the literature describing these techniques is voluminous. After the boom of the 1930's, the war years of the 1940's were a setback to gold mining activity. War Board Order L–208 greatly restricted the development of gold mines; prospecting and mining metals essential to the war effort was deemed more important than mining for gold. More important, however, the economy of the 1940's encouraged work in offices, factories, war industries, and supporting industries, for those not in military service. Many miners and prospectors left the field for the cities and never returned. New Mexico, never a rich gold placer State in comparison with California or Oregon, never again attained the productivity in placer mining it once knew.

PURPOSE AND SCOPE OF PRESENT STUDY

The present paper is a compilation of published information relating to the placer gold deposits of New Mexico, one of a series of four papers describing the gold placer deposits in the Southwestern States. The purpose of the paper is to outline areas of placer deposits in New Mexico and to serve as a guide to their location, extent, production history, and source. The work was undertaken as part of the investigation of the distribution of known gold occurrences in the Western United States.

Each placer is described briefly. Location is given by geographic area and township and range. Topographic maps and geologic maps which show the placer area are listed. Access to each area is indicated by direction and distance along major roads and highways from a nearby center of population.

Detailed information relating to the exact location of placer deposits, their thickness, distribution, and average gold content (all values cited in the text have been converted to gold at $35 per ounce, except where otherwise noted) is included, where available, under the section entitled "Extent."

Discovery of placer gold and subsequent placer-mining activity are briefly described in the section entitled "Production History." Detailed discussion of mining operations is omitted, as this information can be found in the individual papers published by the State of New Mexico, in the yearly Mineral Resources and the Mineral Yearbook volumes published by the U.S. Bureau of Mines and the U.S. Geological Survey, and in many mining journals. Placer gold production, in ounces (table 1) was compiled from the yearly Mineral Resources and Mineral Yearbook volumes and from information supplied by the U.S. Bureau of Mines. These totals of recorded production are probably lower than actual gold production, because substantial amounts of coarse placer gold commonly sold by indivduals to jewelers and specimen buyers are not reported to the U.S. Bureau of Mines or to the U.S. Bureau of Mint. Information about the age and type of lode deposit that was the source of the placer gold is discussed for each district.

A detailed search of the geologic and mining literature was made for information concerning all the placers. A list of literature references is given with each district; the annotation indicates the type of information found in each reference. Sources of information are detailed reports on mining districts, general geologic reports, Federal and State publications, and brief articles and news notes in mining journals. There are five excellent general source books for the mining districts of New Mexico—Jones, 1904; Lindgren, Graton, and Gordon, 1910; Lasky and Wooten, 1933; Anderson, 1957; and Howard, 1967. These source books give the location and history of most placer districts in the State; many other publications give additional information about the placers. A complete bibliography, at the end of the paper, includes separate sections for all literature references and all geologic map references.

Map publications of the Geological Survey can be ordered from the U.S. Geological Survey, Distribution Section, Denver Federal Center, Denver, Colo. 80225; book publications, from the Superintendent of Documents, Government Printing Office, Washington, D.C. 20402.

COLFAX COUNTY

1. ELIZABETHTOWN DISTRICT

Location: Moreno River Valley, west flank of Baldy Mountain, Tps. 27 and 28 N., R. 16 E. (projected; on Maxwell Land Grant).

Topographic maps: Eagle Nest and Red River Pass 7½-minute quadrangles.

Geologic maps:

Bachman and Dane, 1962, Preliminary geologic map of the northeastern part of New Mexico, scale 1:380,160.

Ray and Smith, 1941, Geologic map and structure sections of Moreno Valley (pl. 1), scale 1⅛ in. = 2 miles; Physiographic map and profiles of Moreno Valley (pl. 2).

Access: From Taos, 30 miles northeast on U.S. Highway 64 to Eagle Nest. State Highway 38 leads north 5 miles to Elizabethtown and surrounding placers.

Extent: Placers are found on the slopes of Baldy Mountain, in gulches tributary to the Moreno River from the east, and in the gravels of the Moreno River. Most of the placer mining was concentrated in the area along the lower slopes and along the Moreno River Valley between Anniseta Gulch (2 miles south of Elizabethtown) north to Mills Gulch (3 miles north of Elizabethtown). Some gold was recovered from gulches on the west side of the Moreno River before 1900 (West Moreno; Hematite district).

The sediments in the Moreno River Valley consist of a thick sequence (more than 300 ft. thick) of locally derived unconsolidated sand and gravels that range in age from Pliocene(?) to Holocene. Although the geology of the placer gravels has not been studied in detail, some of these deposits, such as those exposed in deep pits about a quarter of a mile east of Elizabethtown, are believed to be correlative with the Eagle Nest Formation (Pliocene) exposed in the southern part of the Moreno River Valley (Ray and Smith, 1941). The Eagle Nest Formation is considered to be a series of coalescing stream fan deposits which filled the valley during the late Tertiary. The placer gravels on the mountain slopes were generally only a few feet thick and were confined to narrow gulches. Gold, in both the deep river gravels and the shallow slope gravels, was concentrated on the surfaces of hard clay layers, in rich lenses in gravel layers, and in crevices in decomposed bedrock.

Production history: Placer gold valued at more than $3 million was recovered from this district. The Moreno River, Grouse and Humbug Gulches, and Spanish Bar (opposite the mouth of Grouse Gulch) were the most productive placer areas in the district. The greatest part of the gold was recovered during the period 1866–1904. Most of the gold was mined by small-scale sluicing and hydraulic methods, but dredge operations along the Moreno River during the period 1901–5 recovered the major part of gold produced in New Mexico during that period. During this century, most of the placer mining has been on a small scale.

Source: The placer gold in the Elizabethtown district was derived from numerous gold-bearing veinlets which occur in the porphyry bedrock on the west slope of Mount Baldy. This bedrock area has supplied detritus to the Moreno River Valley since the late Tertiary. The veinlets occur throughout the porphyry but in only a few places, as at the Red Bandana group of mines, are they large enough to make lode mining profitable. The quartz-pyrite fissure veins which constitute the Red Bandana lodes are thought to be the principal source of gold in the Grouse Gulch gravels.

Literature:

Anderson, 1956: Production estimates.

———— 1957: General history; production.

Burchard, 1882: Production estimates.

Frost, 1905: Production of El Oro Dredge on Moreno River.

Graton, 1910: Describes extent, value, and origin of placers.

Howard, 1967: Production information for the placers.

Jones, 1903: Describes dredge operations on Moreno River.

———— 1904: Describes history and production; describes placer claims.

Koschmann and Bergendahl, 1968: Production information.

Lasky and Wooton, 1933: Describes thickness of gold-bearing gravels.

Metzger, 1938: Describes problems related to placer mining.

Pettit, 1966a: Ownership history.

———— 1966b: Mining history.

Ray and Smith, 1941: Describes geology of gravels.

Raymond, 1870: Detailed descriptions of placer claims in Grouse and Humbug Gulches.

———— 1872: Production information for 1870.

———— 1873a: Production information for 1871.

———— 1873b: Production information for 1872.

———— 1877: Production information for 1875.

Wells and Wooton, 1932: Extent of gold-bearing gravels.

2. MOUNT BALDY PLACERS

[The Mount Baldy placers here include placers found in the major streams draining the flanks of Baldy Mountain—Willow, Ute, and South Ponil Creeks. Many writers separate the area into small districts named after these major streams]

Location: West, south, and east flanks of Baldy Mountain, T. 27 N., Rs. 16–18 E. (projected; on Maxwell Land Grant).

Topographic map: Ute Park 15-minute quadrangle.

Geologic maps:

  Bachman and Dane, 1962, Preliminary geologic map of the northeastern part of New Mexico, scale 1:380,160.

  Wanek, Read, Robinson, Hays, and McCallum, 1964, Geologic map of the Philmont Ranch region, New Mexico, scale 1:48,000. (See also Robinson and others, 1964.)

Access: From Taos, 42 miles northeast on U.S. Highway 64 to Ute Park and vicinity. Light-duty and dirt roads lead from U.S. Highway 64 to the placer areas in Willow, Ute, and South Ponil Creeks.

Extent: The placers are found in the gravels of Willow, Ute, and South Ponil Creeks. Willow Creek drains the southwest flank of Baldy Mountain; the upper part of the stream flows through a narrow valley, the lower part over a large alluvial fan built up by debris eroded from the mountains by the creek. Most of the placer mining along Willow Creek was concentrated along the upper reaches of the stream, where gold was found in the creek and hillside gravels. Ute Creek drains the southeast flank of Baldy Mountain; placer mining was concentrated in the creek gravels between the Aztec mine downstream to the Atmore Ranch. South Ponil Creek drains the east flank of Baldy Mountain; the placers in this creek were apparently found in the upper reaches near the outcrop of the Aztec vein. Small placers were worked in North Ponil Creek, about 10 miles east of Mount Baldy.

Production history: Placer gold valued at more than $1 million was produced from this area. Most of the production was obtained from Willow and Ute Creeks. Most of the placer mining was on a small scale, although dredges worked on both Willow and Ute Creeks for a few years.

Source: The gold in Willow Creek is derived, at least in part, from small gold-bearing veinlets found in the porphyry bedrock near the northwest fork of Willow Creek. The gold in Ute and South Ponil Creeks is believed to be derived from the eroded outcrops of the Aztec vein, which is located on the ridge separating the two creeks.

Literature:

  Anderson, 1956: Production estimates.

  ———— 1957: Brief outline of placer mining operations; production.

  Burchard, 1882: Production estimates.

  Chase and Muir, 1923: Discovery of placers on Ute Creek.

  Howard, 1967: Production information for placers.

  Jones, 1904: Describes history and production; describes gold in Ute Creek placers.

  Koschmann and Bergendahl, 1968: Production information.

  Lasky and Wooton, 1933: Source of gold in Ponil placers.

  Lee, 1916: Placer discovery.

  Metzger, 1938: Describes placer mining in 1935.

  Pettit, 1966a: Ownership history.

———— 1966b: Mining history.

Raymond, 1870: Details of placer claims in Willow Creek.

———— 1872: Production information for Willow Creek placers for 1870.

Robinson, Wanek, Hays, and McCallum, 1964: Brief description of Ute Creek placers.

3. CIMARRONCITO DISTRICT

Location: South flank of Black Mountain in the Cimarron Range, south of the Cimarron River, T. 26 N., R. 18 E. (projected; on Maxwell Land Grant).

Topographic map: Tooth of Time 15-minute quadrangle.

Geologic maps:

Bachman and Dane, 1962, Preliminary geologic map of the northeastern part of New Mexico, scale 1:380,160.

Wanek, Read, Robinson, Hays, and McCallum, 1964, Geologic map of the Philmont Ranch region, New Mexico, scale 1:48,000.

Access: From Taos, 57 miles northeast on U.S. Highway 64 to Cimarron and junction with State Highway 21; from there, 3 miles south to Philmont Scout Ranch Headquarters. Dirt roads lead to vicinity of Black Mountain and Urraca Creek.

Extent: Placers were reported on Urraca Creek and tributaries.

Production history: Placers were apparently worked in 1898, but no production was recorded.

Source: Unknown.

Literature:

Anderson, 1957: Lists Cimarroncito as placer district.

Jones, 1904: Locates gold-bearing creeks.

Mining Reporter, 1898a: Placer mining in 1898.

GRANT COUNTY

4. WHITE SIGNAL DISTRICT

[White Signal district here includes placer deposits of districts previously named Malone (Gillerman, 1964; Anderson, 1957; Jones, 1904) or southwestern and central Big Burro Mountains (Gillerman, 1964); the deposits are included in the White Signal district by Howard (1967)]

Location: South flank of the Big Burro Mountains, T. 20 S., Rs. 14–16 W.

Topographic maps: Burro Peak and White Signal 7½-minute quadrangles; Redrock 15-minute quadrangle.

Geologic maps:

Ballman, 1960, Geology of the Knight Peak area, scale 1:63,360.

Dane and Bachman, 1961, Preliminary geologic map of the southwestern part of New Mexico, scale 1:380,160.

Gillerman, 1964, Geologic map of western Grant County (pl. 1), scale 1 : 126,720.

Access: From Lordsburg, 29 miles northeast on State Highway 180 to White Signal. Dirt roads lead from main highway at various points to vicinity of different placer localties.

Extent: Placers in the White Signal district are found in two areas: (1) along Gold Gulch and Thompson's Canyon (T. 20 S., R. 16 W., Redrock 15-minute quadrangle; Burro Peak quadrangle) especially in that part of Gold Gulch which traverses secs. 21 and 22 (T. 20 S., R. 16 W., Burro Peak quadrangle; Sunset Goldfields placer) and in a small tributary to Gold Gulch (sec. 27, T. 20 S., R. 16 W., Burro Peak quadrangle; Cureton placer); and (2) in the vicinity of Gold Lake (sec. 20, T. 20 S., R. 14 W., White Signal quadrangle) about 10 miles east of Gold Gulch.

There have been reports of placer gold found in unspecified streams and drywashes within the Burro Mountains. One report indicates that 2 ounces of placer gold was recovered from the Paymaster claim (attributed to the Burro Mountain district in 1942), which may be part of the Paymaster group of lode claims (secs. 21 and 28, T. 20 S., R. 15 W., Burro Peak quadrangle).

Production history: The placers in Thompson's Canyon and Gold Gulch have been worked at least since 1884, and early production is unknown. Most of the activity in this century occurred during the 1930's under the direction of a small placer mining company. The Gold Lake Placer was worked during the period 1900–10 and again during the period 1931–32.

Source: The origin of the gold in Thompson's Canyon and Gold Gulch is unknown, but probably is in gold veins that occur in the adjacent mountains. The gold found in the Gold Lake area was derived from small veinlets in a small knob of granite which protrudes through the alluvium at the lake.

Literature:

Anderson, 1957: Reports placer occurrence.

Ballman, 1960: Describes geology of Gold Gulch area.

Burchard, 1885: Reports that placers in Thompson's Canyon and Gold Gulch produced large amounts of gold.

Gillerman, 1964: Describes placers in White Signal district.

Howard, 1967: Locates placers.

Jones, 1904: Describes placer occurrence.

Raymond, 1877: Reports placers in the Burro Mountains.

U.S. Bureau of Mines, 1941: Reports production of placer gold in the Burro Mountains.

———— 1942: Reports production of placer gold in the Burro Mountains. .

5. PINOS ALTOS DISTRICT

Location: Pinos Altos Mountains, Tps. 16 and 17 S., Rs. 13 and 14 W.

Topographic maps: All 7½-minute quadrangles—Fort Bayard, Reading Mountain, Twin Sisters, Silver City.

Geologic maps:

Dane and Bachman, 1961, Preliminary geologic map of the southwestern part of New Mexico, scale 1:380,160.

Paige, 1911, Geologic relations of fissure veins near Pinos Altos (fig. 10), scale approximately 1 in. = 1 mile.

Access: From Silver City, 8 miles north-northeast on State Highway 25 to the town of Pinos Altos. Dirt roads lead from the town into surrounding hills and placer localities.

Extent: The placers in the Pinos Altos district are found in the vicinity of the sulfide-gold-silver veins in the district, and the gold is commonly concentrated in the gulches below the oxidized vein outcrops. The richest placers were found in Bear Creek (especially near sec. 30, T. 16 S., R. 13 W., Twin Sisters quadrangle), Rich Gulch (near the Mountain Key Mine, in sec. 6, T. 17 S., R. 13 W.; sec. 31, T. 16 S., R. 13 W., Fort Bayard quadrangle), Whiskey Gulch (or Rio de Arenas; there is an uncertainty regarding the position of this gulch on the Fort Bayard quadrangle) and Santo Domingo Gulch (unlocated). Many small gulches which drain near the oxidized vein outcrops were also worked for placer gold.

Production history: The placers were discovered in 1860, and they have been worked practically every year since, mostly by individuals using pans, rockers, and small sluices. Although the richest parts of the placers were probably worked out in the first few years after discovery, many miners continued to work the small gulches during the rainy seasons. Bear Creek and Santo Domingo Gulch were dredged in 1935 and during the period 1939–42, but the location of these operations is not known to me. Production from Santo Domingo Gulch in 1935 was credited to the Central district, but this was apparently an error; I have therefore changed the production table to include all production from Santo Domingo Gulch with Pinos Altos district.

Source: The gold was derived from the eroded outcrops of oxidized sulfide-gold-silver veins in the Pinos Altos district.

Literature:

Anderson, 1957: Locates placers.

Burchard, 1882: History; production.

———— 1883: Describes extent of placer mining.

Bush, 1915: Early history of placer mining.

Graton, Lindgren, and Hill, 1910: Locates placers; mining during the period 1904–5.

Hernon, 1953: Brief history.

Howard, 1967: History; placer mining during the period 1939–41.

Jones, 1904: Describes history and methods of placer mining.

Koschmann and Bergendahl, 1968: Production estimates.

Lasky and Wooton, 1933: Production estimates.

Paige, 1911: Sketch map on page 110 locates the principal lode mines near which the placers were found.

———— 1916: Geology of recent gravels.

Raymond, 1870: Early history of discovery and production of placers.

———— 1872: Production information for 1870.

Schilling, 1959: Road log locates placers.

U.S. Bureau of Mines, 1935: Reports dredge operations on Santo Domingo Creek.

———— 1941: Reports dredge operations on Santo Domingo Creek.

———— 1942: Reports dredge operations on Santo Domingo Creek.

Wolle, 1957: Placer gold recovered in 1955.

Wright, 1915: History.

Wells and Wooton, 1932: Reports black sand analyses.

6. BAYARD AREA

Location: Southeast of the Pinos Altos Mountains, Tps. 17 and 18 S., Rs. 12 and 13 W.

Topographic maps: Fort Bayard and Santa Rita 7½-minute quadrangles.

Geologic maps:

Dane and Bachman, 1961, Preliminary geologic map of the southwestern part of New Mexico, scale 1:380,160.

Lasky, 1936, Geologic map of the Bayard area and outcrops of veins and faults in the Bayard area (pls. 1, 9), scale 1:12,000.

Access: From Silver City, 10 miles east on U.S. Highway 260 to Bayard. Dirt roads lead to mining areas and small placers.

Extent: Placers are found in almost every arroyo that drains the mineralized area in the vicinity of Bayard. The gold is concentrated in workable quantities in only a few areas: (1) south of the Copper Glance vein (secs. 32 and 33, T. 17 S., R. 12 W.; sec. 5, T. 18 S., R. 12 W., Santa Rita quadrangle); (2) the downslope side of the Owl-Dutch Uncle-Tin Box-Lost Mine vein linkage, especially along Gold Gulch (secs. 31 and 32, T. 17 S., R. 12 W.; sec. 5, T. 18 S., R. 12 W.; sec. 6, T. 18 S., R. 12 W., Santa Rita quadrangle), and (3) the vicinity of the veins along Highway 180 between Bayard and Central (sec. 1, T. 18 S., R. 13 W., Fort Bayard quadrangle).

Production history: The Central district is the most productive mining region in New Mexico and includes the areas or subdistricts of Bayard, Santa Rita, Georgetown, and Hanover, where copper, lead, and zinc are mined. The Bayard area is the only part of the Central district with a reported production of placer gold. The small amount of gold credited to the area was recovered by individuals working intermittently over a period of many years.

Source: The gold was derived from the oxidized parts of the quartz-sulfide veins in the district.

Literature:

Frost, 1905: Reports placer mining on Whitewater Creek.

Howard, 1967: Describes source of placer gold.

Lasky, 1936: Describes placers in Bayard area.

HIDALGO COUNTY

7. SYLVANITE SUBDISTRICT

Location: West flank of the south half of the Little Hatchet Mountains, T. 28 S., R. 16 W.

Geologic maps:

Dane and Bachman, 1961, Preliminary geologic map of the southwestern part of New Mexico, scale 1:380,160.

Lasky, 1947, Geologic and topographic map of the Little Hatchet Mountains (pl. 1), scale 1:31,250.

Access: From Lordsburg, 20 miles southeast on U.S. Highway 70–80 to the junction with State Highway 81; from there about 20 miles south to Hachita. Placer ground lies south of Highway 3 (9).

Extent: Placers are found in shallow draws and gulches in gravel remnants on monzonite bedrock between Cottonwood Spring and Livermore Spring (secs. 21 and 28, T. 28 S., R. 16 W.).

Production history: The placers in the Sylvanite subdistrict were discovered in 1908. The discovery created enough excitement to cause a gold rush to the area, but the placers were rapidly worked out. Most production was in 1908, but some placer mining was done in this century at the Bader placer (SW. cor. sec. 21, T. 28 S., R. 16 W.). One report notes that placer gold was discovered at Hachita in 1880.

Source: The gold was derived from the eroded outcrops of the telluride-native gold veins.

Literature:

Anderson, 1957: History.

Dinsmore, 1908: History of discovery.

File and Northrup, 1966; Placer gold at Hachita, 1880.

Hill, 1910: Discovery; extent; production.

Jones, 1908a: Discovery; extent; size of nuggets; production.

———— 1908b: Virtually repeats his article in Engineering and Mining Journal, v. 86, 1908.

Lasky, 1947: Location; extent; geology; age and origin of placer gravels; source of gold; fineness and size of gold particles; placer-mining operations.

LINCOLN COUNTY

8. JICARILLA DISTRICT

Location: Jicarilla Mountains, T. 5 S., R. 12 E.

Topographic map: Roswell 2-degree sheet, Army Map Service.

Geologic maps:

Dane and Bachman, 1958, Preliminary geologic map of the southeastern part of New Mexico, scale 1:380,160.

Griswold, 1959, Geologic map of Lincoln County (pl. 2), scale 1 in. =6 miles; Generalized geologic map of Jicarilla district (fig. 21), scale approximately 1 in. = 1 mile.

Access: Three miles north of Carrizozo, a light-duty road leads east from U.S. Highway 54 through White Oaks Canyon and through the Jicarilla Mountains.

Extent: Placers are found in gulches in the vicinity of the village of Jicarilla. Most of the placers are concentrated in Ancho, Warner, Spring, and Rico Gulches. The area is not mapped on a scale larger than 1:250,000, and the placer gulches cannot be accurately located on that scale.

Production history: The placers in the Jicarilla Mountains, discovered in 1850, have been worked on a small scale, by individuals, for more than 100 years. Although large-scale operations have not been successful because of scarcity of water and depth of overburden, mining has been profitable to the men who worked them on a small scale. From many accounts, it seems that the major part of the placer ground is unworked; however, no detailed studies of the placers have been made, and the extent of any placer ground actually remaining is unknown.

Source: The gold in the gravels is derived from small gold-pyrite veins within the monzonite porphyry intrusion, which forms the Jicarilla Mountains. In places, the gold is found directly above decomposed gold-bearing bedrock.

Literature:

Anderson, 1957: Describes extent of placers; names placer gulches; fineness of gold.

Burchard, 1883: History.

File, 1965: Lists "Rico lease" placer active in 1965.

Graton, 1910a: History; describes thickness of pay gravel and overburden in Ancho Creek placer-mining operations.

Griswold, 1959; Location; extent; names placer gulches; production; source; accessory minerals.

Jones, 1904: Placer-mining techniques; placer-mining operations.

Lasky and Wooton, 1933: Production estimates.

Raymond, 1870: Placer-mining techniques.

Smith and Dominian, 1904: States Spaniards mined Jicarilla placers (about 1700).

Wells and Wooton, 1932: Geology of placers; source; production estimates.

Wright, 1932: Gold values in gravels estimated.

9. WHITE OAKS DISTRICT

Location: In the vicinity of Baxter and Lone Mountains on the west flank of the Jicarilla Range, T. 6 S., R. 11 E.

Topographic maps: Roswell 2-degree sheet, Army Map Service; Little Black Peak and Carrizozo 15-minute quadrangles.

Geologic maps:

Dane and Bachman, 1958, Preliminary geologic map of the southeastern part of New Mexico, scale 1 : 380,160.

Griswold, 1959, Geologic map of Lincoln County (pl. 2), scale 1 in. = 6 miles; Geologic map of a part of the Lone Mountain area (fig. 2).

Smith and Budding, 1959, Little Black Peak, east half, scale 1 : 62,500.

Access: 3 miles north of Carrizozo, a light-duty road leads northeast from U.S. Highway 54 through White Oaks Canyon about 8 miles to placer area.

Extent: Small placers occur in Baxter Gulch and White Oaks Gulch (secs. 35 and 36, T. 6 S., R. 11 E., projected). Placers were also found in small tributaries to these gulches in the vicinity of the lode mines in the district.

Production history: The White Oaks district has been predominantly a lode mining district; the placers were important only to individuals before the discovery of the lode in 1879.

Source: The placer gold was evidently derived from the gold-bearing quartz-pyrite veins of the district, as much of the placer mining was conducted in the vicinity of the major lodes.

Literature:

Anderson, 1957: History; location; placer-mining operations.

Graton, 1910a: History.

Griswold, 1959: Location; placer-mining problems.

Jones, 1904: History.

Smith and Dominian, 1904: Discovery of placers; includes photographs of placer area.

10. NOGAL DISTRICT

Location: Eastern side of the Sierra Blanca Mountains, southwest of Nogal, T. 9 S., Rs. 12 and 13 E.

Topographic map: Capitan 15-minute quadrangle.

Geologic maps:
>   Dane and Bachman, 1958, Preliminary geologic map of the southeastern part of New Mexico, scale 1 : 380,160.
>   Griswold, 1959, Geologic map of Lincoln County (pl. 2).

Access: Eight miles east of Carrizozo on U.S. Highway 380, State Highway 37 leads 4 miles southeast to Nogal. Dirt roads lead from Nogal west to the placer area.

Extent: Placers are found in Dry Gulch, which drains northeast from the Sierra Blanca towards Nogal. The gold-bearing gravels are found about 1 mile below the outcrops of the ore veins (probably sec. 7, T. 9 S., R. 13 E.). Placers are also found at the Dugan-Dixon claim (unlocated).

Production history: The placers have been worked since 1865, but early production is unknown and probably is small. Minor amounts of gold have been recovered by sluicing along Dry Gulch during this century.

Source: The placers are found below the outcrops of the gold-sulfide fissure veins, mined at the Helen Rae and American lodes, and were probably derived from these veins or similar small gold-bearing veins.

Literature:
>   Anderson, 1957: Names placer claims.
>   Burchard, 1885: Production information for 1884.
>   Graton, 1910a: Notes discovery of placers.
>   Griswold, 1959: Placer-mining history; placer-mining operations (1957–58).
>   Jones, 1904: Notes placer discovery.
>   Raymond, 1870: Production information for 1869.

MORA COUNTY

11. MORA RIVER PLACERS (RIO LA CASA DISTRICT)

Location: Mora River Valley, in the western part of Mora County, T. 21 N., R. 15 E. (projected; on the Mora Grant).

Topographic map: Santa Fe 2-degree sheet, Army Map Service.

Geologic map: Bachman and Dane, 1962, Preliminary geologic map of the northeastern part of New Mexico, scale 1 : 380,160.

Access: From Taos, about 40 miles south and southeast to Cleveland on State Highway 3.

Extent: Small amounts of placer gold have been found in the mountain gulches and in old terraces along the Mora River, near the village of Cleveland.

Production history: The gravels in terraces along the Mora River were placered before 1940, but the amount of gold recovered was small. No production has been recorded from this area, and all available information indicates that the Mora River placers are low in tonnage and gold content.

Source: The gold is thought to be derived from numerous quartz lenses and veinlets found near the headwaters of the Rio La Casa and Lujon Creek, 9 miles west of Mora.

Literature:

Anderson, 1957: Location; history; origin.

Harley, 1940: Location; placer-mining operations; size and fineness of gold; source.

Howard, 1967: Locates placers.

Raymond, 1870: Notes presence of placers.

OTERO COUNTY

12. OROGRANDE (JARILLA) DISTRICT

Location: Jarilla Mountains in the Tularosa Valley southeast of the White Sands National Monument, T. 22 S., R. 8 E.

Topographic maps: Orogrande North and Elephant Mountain 7½-minute quadrangles; Orogrande 15-minute quadrangle.

Geologic map: Schmidt and Craddock, 1964, Geologic map of the Jarilla Mountains (pl. 1), scale 1:24,000.

Access: From El Paso, Tex., 45 miles north on U.S. Highway 54 to Orogrande; from Las Cruces, 17 miles northeast on State Highway 3; from there 34 miles east on light-duty road to Orogrande. Dirt roads lead from Orogrande 1½ miles north to the placer area.

Extent: Placer mining was concentrated in the gravels in the NW¼ sec. 14 and NE¼ sec. 15, T. 22 S., R. 8 E. (Orogrande North quadrangle), at the south flank of Jarilla Mountains. Similar environments favorable for the concentration of placer gold occur in secs. 2, 3, 5, 9, 11, 14, and 15, T. 22 S., R. 8 E.

Production history: The Orogrande placers were actively worked in the early part of this century and again during the 1930's. Most of the work was done by individuals who dug many small holes and tunnels in the caliche cemented gravel to follow pay streaks. Most of the gold at the Little Joe claim (NE¼, NW¼ sec. 14, T. 22 S., R. 8 E.) was found in the 6–9 in. of gravel overlying bedrock.

Source: The gold in these placers was derived from complex sulfide ores formed during the closing stages of consolidation of the Tertiary monzonite adamellite, the principal rock of the district.

Literature:

Anderson, 1957: Production estimates; characteristics and values of gold in the gravels.

Gifford, 1899: Placer mining in 1899; extent of placers; size of gold particles; grade of gravels (probably exaggerated).

Graton, 1910b: Location; origin.

Howard, 1967: Origin of the placers.

Jones, 1904: Geology; placer-mining operations; size and value of gold gravels.

Lasky and Wooton, 1933: Production estimates; fineness of gold; gravel values.

Schmidt and Craddock, 1964: Detailed description of placer deposits; includes suggestions for future prospecting; extent; value of gravels; describes gravels in placer pit; distribution of gold in the gravels.

Wells and Wooton, 1932: Black sand analyses.

RIO ARRIBA COUNTY

13. HOPEWELL DISTRICT

Location: Tusas Mountains, in the valley between Jawbone and Burned Mountains southwest of Hopewell Lake, T. 29 N., R. 7 E., secs. 31 and 32; T. 28 N., R. 6 E., secs. 1 and 12.

Topographic maps: Burned Mountain 7½-minute quadrangle; Cebolla 15-minute quadrangle.

Geologic map: Bingler, 1968a, Geologic map of the Hopewell mining district (pl. 4), scale ¾ in. = 1 mile.

Access: From Santa Fe, 72 miles north on U.S. Highway 285 to Tres Piedras; from there about 15 miles east on light-duty road to Hopewell Lake and placer area.

Extent: Placers in the Hopewell district occur in gravels along Placer Creek from Hopewell Lake (sec. 32, T. 29 N., R. 7 E.) probably as far downstream as the mouth of Placer Creek (sec. 12, T. 28 N., R. 6 E.). The Fairview placer, the most extensively mined deposit, is located in the upper flat area, south of Hopewell Lake. Minor placers occur in Placer Creek Gorge and the lower flat area. Placer Creek was formerly known as Eureka Creek. Most of the area is now a campground in the Carson National Forest.

Production history: Most production occurred during the first few years after the discovery, about 1880. Gold valued at more than $175,000 was recovered during the first 3 years; total production to 1910 is estimated to be about $300,000. Mining during the 1900's has been on a small scale.

Source: The gold is derived from gold-bearing sulfide replacement veins and gold-quartz veins found in Precambrian schists and gneisses in the district. The Fairview placer formed during the early Tertiary and is now exhumed by Placer Creek. The placer gravel along Placer Creek and in the Lower Flat area is of recent alluvial origin, but some of the gold may be derived from the Fairview Placer.

Literature:

Anderson, 1957: Production history.

Benjovsky, 1945: Indicates areas with future mining potential.

Bingler, 1968: Location; extent; placer-mining operations; emphasis is on origin of placer gravels and age of deposition.

Burchard, 1884: Placer-mining operations.

Graton, 1910c: Location; history; early production; size of nuggets; depth of gravels; origin of gold.

Howard, 1967: Locates placers.

Jones, 1904: History; early placer operations; origin.

Just, 1937: Reports potential placer at the mouth of Placer Creek.

Wells and Wooton, 1932: History, production, and extent of placers; size of nuggets; depth of gravels.

14. EL RITO REGION

Location: Southwest of Valle Grande Peak in the Chama Basin, T. 25 N., R. 7 E.

Topographic maps: El Rito and Valle Grande 7½-minute quadrangles.

Geologic maps:

Bingler, 1968b, Reconnaissance geology of the El Rito 7½-minute quadrangle, scale 1:24,000.

———— 1968c, Reconnaissance geology of the Valle Grande Peak 7½-minute quadrangle, scale 1:24,000.

Access: El Rito is on State Highway 96, 12 miles north of U.S. Highway 84. The placer area is accessible by dirt road from El Rito.

Extent: Small quantities of gold occur in the conglomerate found between the drainages of El Rito Creek and Arroyo Seco.

Production history: There is no reported production from this area. The gold content of the conglomerate assays about 10¢ per ton; no commercial importance is attached to the occurrence.

Source: Unknown.

Literature:

Bingler, 1968a: Reports no evidence of prospecting or mining activity in the El Rito district.

Howard, 1967: Locates deposits.

Lasky and Wooton, 1933: Reports noncommercial concentrations of gold in the conglomerates of the Santa Fe Formation.

Wells and Wooton, 1932: Describes placer occurrence; promotion activities and assay values.

15. RIO CHAMA PLACER (ABIQUIU DISTRICT)

Location: In the valley of the Rio Chama, a few miles upstream from Abiquiu, T. 23 N., Rs. 5 and 6 E. (projected; on Juan Jose Lobato Grant).

Topographic map: Abiquiu 15-minute quadrangle.

Geologic map: Bingler, 1968a, Geologic map of Rio Arriba County, east half (pl. 1b), scale approximately ½ in. = 1 mile.

Access: The Rio Chama is accessible by dirt roads leading west from Abiquiu.

Extent: Gold reportedly occurs in the river sands and gravels along the Rio Chama a few miles above Abiquiu, but the exact location and extent of these deposits is unknown.

Production history: Placers have been worked along the Rio Chama near Abiquiu, possibly before 1848 and during the 1880's to 1900's. These deposits received some attention from mining companies before 1900, and some large-scale operations were attempted. News reports indicate that the placers were productive, although no production records are known. The district has apparently remained unworked throughout most of this century.

Source: Unknown, but thought to be derived from low-grade gold deposits in Precambrian rocks exposed along the deep canyon of the Rio Chama.

Literature:

Anderson, 1957: Location.

Bancroft, 1889: Reports knowledge of gold before 1848 near Abiquiu.

Burchard, 1885: Placer-mining activity.

Graton, 1910c: Reports placer occurrence; origin.

Howard, 1967: Reports placer occurrence.

Jones, 1904: Extent; placer-mining operations; thickness of gravels; gold values in bench gravels and river gravel.

Mining Reporter, 1898a: Placer-mining operations.

Prince, 1883: Reports placers known in 1844.

SANDOVAL, BERNALILLO, AND VALENCIA COUNTIES

[Small placer deposits occur in three widely separated localities on the flanks of the Sandia Mountains, which trend north-south through Sandoval, Bernalillo, and Valencia Counties. The presence of placers has been known for probably two centuries; many legends state that Spaniards engaged in mining in the area and probably prospected for placer gold. Despite the long history of the area, little is known of the exact location, extent, or worth of these placer deposits]

16. PLACITAS—TEJON REGION

Location: North end of the Sandia Mountains, T. 13 N., R. 5 E. (on the San Antonio de las Huertas Grant and the town of Tejon Grant).

Topographic map: San Felipe Pueblo 15-minute quadrangle.

Geologic map: Dane and Bachman, 1957, Preliminary geologic map of the northwestern part of New Mexico, scale 1 : 380,160.

Access: From Albuquerque, 17 miles north on U.S. Interstate 25 to State Highway 44; from there, about 7 miles east to Placitas (Sandoval County).

Extent: Placers are found in the area around the towns of Placitas and Tejon, at the north end of the Sandia Mountains in the vicinity of Las

Huertas Creek and Tejon Canyon (center sections of T. 13 N., R. 5 E.).

Production history: Gold reportedly occurs in beds of cemented gravels in this region; individuals using drywashing machines reportedly recovered gold worth $3 per day per man during the first decade of this century.

Source: Unknown.

Literature:

Anderson, 1957: Mentions presence of placers.

Elston, 1967: Describes vein mineralization; no placer information.

Heikes and York, 1913: Placer-mining operations; production; type of gravels.

Jones, 1904: Reports occurrence of auriferous cemented gravels.

Wells and Wooton, 1932: Production information for 1904.

17. TIJERAS CANYON REGION

Location: Central part of the Sandia Mountains, T. 10 N., Rs. 4–6 E.

Topographic maps: Albuquerque 15-minute quadrangle; Tijeras 7½-minute quadrangle.

Geologic map: Dane and Bachman, 1957, Preliminary geologic map of the northwestern part of New Mexico, scale 1 : 380,160.

Access: From Albuquerque, about 18 miles east on U.S. Highway 66 to Tijeras Canyon (Bernalillo County).

Extent: The dry streams around Tijeras Canyon and the alluvial flats between the Sandia Mountains and Albuquerque on the west have been drywashed intermittently.

Production history: No production has been recorded from this area. The area is now included in the Sandia Military base and is not accessible to prospectors.

Source: The gold was probably derived from small quartz lenses in Tijeras Canyon formed during the Precambrian mineralization which contain native gold in the oxidized parts of the veins.

Literature:

Anderson, 1957: Lists as placer district; no description.

Burchard, 1882: Reports placer excitement in Tijeras Canyon area in 1881.

———— 1884: Reports placer excitement in the Rio Grande north of Albuquerque in 1883; size of gold nuggets found.

Elston, 1967: Describes bedrock mineralization; no placer descriptions.

Howard, 1967: Placer prospecting history.

18. HELL CANYON REGION

Location: Southern part of the Sandia Mountains, T. 8 N., Rs· 3–5 E. (on the Isleta Pueblo Grant).

Topographic map: Mount Washington 7½-minute quadrangle.

Geologic maps:

Dane and Bachman, 1957, Preliminary geologic map of the northwestern part of New Mexico, scale 1:380,160.

Reiche, 1949, Geologic map of the Manzanita and North Manzano Mountains (pl. 5), scale approximately 1 in. = 1 mile.

Access: From Albuquerque about 13 miles south on U.S. Interstate 25 to Isleta; from there dirt roads lead east along Hell Canyon (Valencia County).

Extent: Placers are found in the gravels in Hell Canyon and other drywashes in that vicinity. A narrow strip of placer ground is located at the west end of the Milagras group of patented claims (sec. 29, T. 8 N., R. 5 E.).

Production history: No recorded placer production.

Source: The gold was probably derived from the gold sulfide ores mined at the Milagras group of lode claims.

Literature:

Jones, 1904: Reports placer occurrence; placer-mining development.

Reiche, 1949: Locates placer claim in Hell Canyon; describes lodes.

SAN MIGUEL COUNTY

19. WILLOW CREEK DISTRICT

Location: East side of the Pecos River in the Sangre de Cristo Range, T. 18 N., Rs. 12 and 13 E.

Topographic map: Cowles 7½-minute quadrangle.

Geologic maps:

Bachman and Dane, 1962, Preliminary geologic map of the northeastern part of New Mexico, scale 1:380,160.

Miller, Montgomery, and Sutherland, 1963, Geology of part of the southern Sangre de Cristo Mountains, New Mexico (pl. 1), scale 1:63,360.

Access: From Pecos about 12 miles north to junction of Willow Creek and the Pecos River.

Extent: Willow Creek is frequently listed as a placer locality, but no description of the deposits has been found in the literature. The Pecos Copper mine is located at the junction of Willow Creek and the Pecos River; the ores contain a complex mixture of sulfides, gold, and silver. It is possible that some placer gold was recovered from debris eroded from this lode.

Production history: The presence of placer gold in this area has been known for many years, but there is no reported placer production.

Source: Unknown.

Literature:

Anderson, 1957: Lists as placer district; no description.

Burchard, 1883: Reports presence of gold in streams draining Sangre de Cristo Mountains.

Lasky and Wooton, 1933: Lists as placer district; no description.

20. VILLANUEVA AREA

Location: Northeast side of the Pecos River between Villanueva and Sena, T. 12 N., R. 15 E.

Topographic map: Villanueva 15-minute quadrangle.

Geologic map: Johnson, 1970, Geologic map of the Villanueva quadrangle, scale 1:62,500.

Access: Villanueva is located on State Highway 3, 12 miles south of U.S. Highway 85.

Extent: A fossil placer occurs within crossbedded layers in the Permian sandstone (probably Yeso Formation, but referred to as Glorieta Formation by Harley, 1940) exposed in bold cliffs along the Pecos River between Sena and Villnueva (NE¼ T. 12 N., R. 15 E.). The placers were prospected before 1940 by three tunnels driven into lenses of crossbedded sandstone between layers of thin impervious shaly sandstone; small amounts of gold are concentrated in small pockets within the lenses.

Production history: Assays made on samples from the sandstone show variable amounts of gold—as high as $22.00 per ton and as low as a trace of gold; the deposit is considered uneconomic.

Source: Unknown. The Precambrian rocks exposed north of Villanueva might contain a small amount of gold in quartz veins and might have been the source of this deposit.

Literature:

Harley, 1940: Location; extent; prospecting activity; distribution of gold in sandstone; origin; assays.

21. LAS VEGAS AREA

Location: Within the city of Las Vegas, T. 16 N., R. 16 E. (projected; within the Las Vegas Grant).

Topographic map: Las Vegas 7½-minute quadrangle.

Geologic map: Bachman and Dane, 1962, Preliminary geologic maps of the northeastern part of New Mexico, scale 1:380,160.

Access: From Santa Fe, about 55 miles south and east on U.S. Highway 85–84 to Las Vegas.

Extent: In 1883, placer gold was discovered during building excavation for the courthouse within the city of Las Vegas. A brief excitement followed, during which many placer claims were staked around the town within a radius of 2 miles.

Production history: No production has been recorded.

Source: Unknown.

Literature:

Burchard, 1884: Location; placer-mining activity; results of test samples.

SANTA FE COUNTY

22. OLD PLACERS DISTRICT (DOLORES, ORTIZ)

Location: Ortiz Mountains, Tps. 12 and 13 N., Rs. 7 and 8 E. (projected: on the Ortiz Mine Grant).

Topographic map: Madrid 15-minute quadrangle.

Geologic map: Dane and Bachman, 1957, Preliminary geologic map of the northwestern part of New Mexico, scale 1:380,160.

Access: Placers are accessible by dirt roads leading east from State Highway 10 near Cerrillos (north side of mountains) or near Golden (south side of mountains).

Extent: Placers are found on the eastern and southern slopes of the Ortiz Mountains; the gold occurs in both creek and mesa gravels. The richest gravels are found on the eastern slope of the mountains in the vicinity of Dolores and Cunningham Gulch; here, the gold is in mesa gravels that are the upper part of an old debris fan formed at the mouth of Cunningham Gulch. Substantial amounts of gold were also recovered from creek gravels in Dolores Gulch and Arroyo Viejo. Placers are also found on the southern slope of the mountains, north of Arroyo Tuerto, in the vicinity of Lucas Canyon; these deposits were not so rich as those found near the town of Dolores.

Production history: The placers in the Ortiz Mountains were discovered in 1828; they have been worked on a small scale since that time. Because the gold is concentrated in pay streaks that have sporadic distribution, large-scale operations have not been successful, and the individual with the small hand-carried drywashing machine was more effective in mining the gravels than were the large companies with less mobile gold-concentrating plants. Despite the problems encountered in mining the placers, the deposits in the Ortiz Mountains have produced placer gold worth about $2 million.

Source: The placers in the district were derived from the local veins. In Cunningham Gulch, near the richest placers, two important types of ore deposits occur in brecciated margins of a trachyte-latite vent rock: gold in distinct quartz fissure veins and stringers (for example, the Ortiz mine); disseminated gold and scheelite (for example, Cunningham mine). On the south side of the mountains, contact pyrometasomatic auriferous pyrite, chalcopyrite, and scheelite are disseminated in garnet tactite (for example, Lucas and Candelaria mines—no significant production); the placers here are less rich.

Literature:

Anderson, 1957: Virtually repeats Lasky and Wooton (1933); placer mining operations after World War II.

Blake, 1859: Early mining practices.

Burchard, 1883: History; production information.

Elston, 1967: Location; repeats description by Lindgren (1910); placer mining operations during the period 1939–40 (see Smith, 1940).

Engineering and Mining Journal, 1899: Average value of gravels.

Harrington, 1939: Describes discovery; mining techniques; problems.

Hartly, 1915: Placer-mining problems at Cunningham Mesa.

Howard, 1967: Locates placer claims.

Jones, 1904: Detailed history and early production information.

Koschmann and Bergendahl, 1968: Location; extent; production.

Lasky and Wooton, 1933: Characteristics of placer gravels; gold values in black sand and concentrates.

Lindgren, 1910: Location; extent; depth of gravels.

Prince, 1883: Production from 1832 to 1835, and following.

Raymond, 1870: Location; early mining practices.

———— 1874: Raymond completed a detailed study of the Ortiz mine grant in 1873; he publishes the results of his studies on both lodes and placers in this report. The details of the extent of gold-bearing gravels and early placer-mining operations are described.

Smith, 1940: Describes large operation on placers near Dolores.

23. NEW PLACERS DISTRICT (SAN PEDRO)

Location: San Pedro Mountains, T. 12 N., R. 7 E.,

Topographic maps: San Pedro 7½-minute quadrangle; Edgewood and Madrid 15-minute quadrangles.

Geologic maps:

Atkinson, 1961, Geologic map of the San Pedro Mountains (pl. 4), scale 4 in.=1 mile.

Dane and Bachman, 1957, Preliminary geologic map of the northwestern part of New Mexico, scale 1:380,160.

Access: Placers are accessible by dirt roads leading east from State Highway 10 near Golden.

Extent: Placers are found on the north, south, and west flanks of the San Pedro Mountains; the gold is found in subangular detritus at the foot of the mountains and has been further concentrated in creeks and gulches that cut into these gravel beds. The richest gravels were found on the north side of the mountains where the gold was recovered from gravels in branches of the Arroyo Tuerto near Golden (especially in Old Timer Creek in secs. 17–20, T. 12 N., R. 7 E., Madrid quadrangle) and on the south side of the mountains where the gold was recovered from gravels in San Lazarus Creek (sec. 27, T. 12 N., R. 7 E., San Pedro quadrangle). Cemented gravels in the vicinity of Golden received considerable attention during the first part of this century, and many plans were made to extract the gold presumed to be contained in the gravels. All attempts to mine these gravels have failed.

Production history: The placers in the San Pedro Mountains were discovered in 1839; like the placers in the Ortiz Mountains, they have been worked on a small scale since that time. Large-scale operations have always been inhibited because of lack of water for wet-concentration methods, and the gravels are too wet for dry-concentration methods. Despite these difficulties, the New Placers district is credited with a production of placer gold valued at nearly $2 million.

Source: The placers were derived by erosion of (1) small shear zones and fissures in intrusive rocks of early Oligocene-to-early Miocene age; filled with quartz and auriferous pyrite containing free gold in the oxidized zone, and (2) small pockets of auriferous pyrite disseminated in tactite, containing free gold in the oxidized zone.

Literature:

Anderson, 1957: Virtually repeats information given in Lasky and Wooton; adds production information for 1931–52.

Atkinson, 1961: Placers were not examined in detail in the course of this study, and the description of the deposits is taken from earlier writers; adds commentary about production.

Blake, 1859: Early mining practices; describes nuggets.

Brinsmade, 1908: Placer-mining operations; characteristics of the gravels.

Burchard, 1883: History; production.

———— 1884: Placer-mining operations; average value of gravels.

———— 1885: Placer-mining operations; production.

Elston, 1967: Location; extent, production, and origin of the placers; average gold values in gravels; placer-mining operations.

File, 1965: Lists active placer claims.

Harrington, 1939: History; early placer-mining techniques.

Herrick, 1898: Origin of the placers.

Howard, 1967: Locates placer claims.

James, 1955: History; reports observations of a visit in 1939.

Jones, 1904: Outlines placer area and production.

———— 1906: Report on value of "cement gravels."

Lasky and Wooton, 1933: Location, extent, and origin of placer. Gold values in gravel.

Lindgren, 1910: History; placer-mining operations; extent; geology.

Koschmann and Bergendahl, 1968: Production information.

Prince, 1883: History; production information to 1845.

Raymond, 1870: Early placer-mining techniques.

Statz, 1912: Discusses origin of placers.

Yung and McCaffrey, 1903: Extent and value of gravels.

24. SANTA FE DISTRICT

Location: Sangre de Cristo Mountains, T. 17 N., R. 11 E.

Topographic map: Aspen Basin 7½-minute quadrangle.

Geologic map: Bachman and Dane, 1962, Preliminary geologic map of the northeastern part of New Mexico, scale 1 : 380,160.

Access: From Santa Fe, a dirt road leads up Santa Fe River to McClure Reservoir; from there, apparently no roads or trails exist.

Extent: Placer gold has been reported to occur in the upper reaches of the Santa Fe River.

Production history: No recorded production.

Source: Unknown.

Literature:

Anderson, 1957: Notes placer occurrence.

Burchard, 1884: Notes placer occurrence; location.

Elston, 1967: Notes placer occurrence and prospecting activity.

Lasky and Wooton, 1933: Notes placer occurrence.

SIERRA COUNTY

25. HILLSBORO DISTRICT (LAS ANIMAS PLACERS)

Location: Animas Hills, in the eastern foothills of the Black Range, Tps. 15 and 16 S., Rs. 6 and 7 W.

Topographic maps: Hillsboro 15-minute quadrangle; Skute Stone Arroyo 7½-minute quadrangle.

Geologic maps:

Dane and Bachman, 1961, Preliminary geologic map of the southwestern part of New Mexico, scale 1 : 380,160.

Harley, 1934, Topographic and geologic map of the Hillsboro (Las Animas) lode mining district (pl. 6), scale about 2½ in. = 4,000 ft; general map of the Hillsboro (Las Animas) placer mining district (pl. 7), scale approximately 1 in. = 1 mile.

Kuellmer, F. J., compiler, 1956, Geologic map of Hillsboro Peak, 30-minute quadrangle, scale 1 : 126,720.

Access: Placers are accessible by dirt roads leading north from State Highway 90, 5 miles east of Hillsboro and 13 miles west of Interstate 25.

Extent: Placers are found on the east and south flanks of the Animas Hills. The placers on the east flank occupy a larger area roughly bounded by Dutch Gulch on the north and the Rio Percha on the south (in Tps. 15 and 16 S., Rs. 6 and 7 W.). The placers on the south flank are found in Snake Gulch (secs. 3 and 4, T. 16 S., R. 7 W.) and Wicks Gulch (secs. 1 and 2, T. 16 S., R. 7 W.).

The major placers are concentrated on the east flank in area drained by Dutch, Grayback, Hunkidori, Greenhorn, Gold Run, and Little Gold

Run Gulches. These areas are shown on the Hillsboro quadrangle in secs. 25 and 36, T. 15 S., R. 6 W., and secs. 4–6 and 7–18, T. 16 S., R. 6 W. Surface samples of alluvium contain fine gold for 3 miles east of the junction of Dutch and Hunkidori Gulches (NE¼ sec. 33, T. 15 S., R. 6 W.).

Production history: The placers on the east flank of the Animas Hills produced placer gold valued at about $2,060,000 between 1877 and 1931. Between 1934 and 1937, large-scale placer-mining operations worked the ground along Gold Run Gulch; the success of these operations has credited the Hillsboro district with the largest amount of placer gold recovered during this century in New Mexico. The deposits in Snake and Wicks Gulches were mined by hand methods, mostly during the years immediately following the placer discovery in 1877. These deposits produced gold worth about $140,000, of which $90,000 was recovered from Wicks Gulch during the winter of 1877–78.

Source: The Tertiary (Oligocene?) andesites exposed in the Animas Hills contain gold-bearing fissure veins which are the principal source of lode gold in the district; erosion of parts of the andesite has concentrated the gold in the intermediate part of the alluvial fans on the east flank of the Animas Hills. Erosion of similar small veins has concentrated gold in Snake and Wicks Gulches on the south flank of the hills.

Literature:

Anderson, 1957: Placer-mining operations during the period 1935–42.

Burchard, 1882: Placer-mining problems; production.

——— 1883: Placer-mining operations at Snake Gulch.

——— 1885: Production information for 1884.

Endlich, 1883: Source of placer gold; placer mining in 1883.

Gardner and Allsman, 1938: Placer-mining operations with movable plant; depth of gravels; clay content.

Harley, 1934: Gives a detailed description of the placers in the Hillsboro district. History, extent, geology, origin, and production of placers.

Heikes and Yale, 1913: Extent and value of placers; thickness of gravels.

Howard, 1967: Virtually repeats description of Harley (1934); notes importance of placer mining in the district.

Jones, 1904: History; early production.

Koschmann and Bergendahl, 1968: Production.

Leatherbee, 1911: History; extent; production; mining operations.

26. PITTSBURG DISTRICT (SHANDON, SIERRA CABALLO)

Location: West flank of the Caballo Mountains adjacent to the Caballo Reservoir, T. 16 S., R. 4 W.; T. 14 S., R. 4 W.; T. 17 S., R. 4 W.

Topographic maps: All 7½-minute quadrangles—Caballo, Williamsburg, Garfield.

Geologic maps:

Dane and Bachman, 1961, Preliminary geologic map of the southwestern part of New Mexico, scale 1:380,160.

Harley, 1934, General map of the Pittsburg (Shandon) placer mining district (fig. 19), scale approximately 1 in. = 1 mile.

Kelley and Silver, 1952, Geologic map of the Caballo Mountains (fig. 2), scale approximately 1 in. = 1 mile.

Access: From Las Cruces, 56 miles north on Interstate 25 to Caballo Dam Road; from there, 1 mile east to dirt roads that lead northeast about 1 mile to the placer area.

Extent: Placers are found in the alluvial fan near the base of the escarpment of the Caballo Mountains, in a small area in Trujillo Gulch and the area drained by its tributaries, and in Apache Canyon and Union Gulch (secs. 16 and 17, 20–22, T. 16 S., R. 4 W., Caballo quadrangle). (Trujillo Gulch is named "Caballo Canyon" on the topographic map.) Small placers reportedly occur north of the Pittsburg placer area in gulches west of Palomas Gap (T. 14 S., R. 4 W., Williamsburg quadrangle), and near Derry, south of Caballo Dam (T. 17 S., R. 4 W., Garfield quadrangle). Other than general location, I have found no information relating to the small placers at Palomas Gap and Derry.

Production history: The Pittsburg placers were discovered in 1901 by Encarnacion Silva, who attempted to keep the location secret. Until 1903, he worked the placers alone, but in November 1903, he revealed the secret at Hillsboro; the news immediately started a gold rush to the area. Many men and some small placer companies worked the gravels almost continuously from 1904 to 1941.

During the period 1932–38 placer miners recovered 32.27 ounces of gold from the Caballo Mountains, and, as this production was recorded separately from that credited to the Pittsburg district, it would appear that the other small placers were worked that year.

Source: The gold in the Pittsburg placers were derived from gold-bearing quartz veins in the Precambrian granites and schists exposed in the lower part of the escarpment; the actual source of gold is believed to have been eroded as the veins now exposed do not show large concentrations of gold.

Literature:

Anderson, 1955: Location; notes placer mining at Palomas Gap.

———— 1957: Placer operations during the period 1933–42 at Pittsburg district; lists Derry as placer district.

Gordon, 1910; Extent; geology; origin.

Harley, 1934: Extent; geology; origin.

Howard, 1967: Reports placer reserves.

Jones, 1904: History (personal account).

Kelley, 1951: Reports occurrence of placer gold in the Cambrian Bliss Sandstone.

Kelley and Silver, 1952: Describes stratigraphy of the placer deposits.
Keyes, 1903: History; geology.

27. CHLORIDE DISTRICT (UPPER CUCHILLO NEGRO PLACERS)

Location: East flank of Lookout Mountain in the Black Range, along the
upper part of the Cuchillo Negro, T. 10 or 11 S., R. 9 W.

Topographic map: Lookout Mountain 15-minute quadrangle.

Geologic map: Dane and Bachman, 1961, Preliminary geologic map of the
southwestern part of New Mexico, scale 1:380,160.

Access: Chloride is on State Highway 52, 29 miles west of U.S. Highway 85,
9 miles north of Truth or Consequences. Dirt roads lead from the town
to surrounding areas.

Extent: Placers are reported to occur on the upper Cuchillo Negro, near
Chloride. Although these deposits were discovered in 1883 and were said
to have been known long before that time, I have found no information
that describes the occurrence in much detail. In 1883 the bars of the
stream were washed and 2–7 colors per pan were recovered.

Production history: The only recorded placer production was 1.79 ounces of
placer gold attributed to the Chloride district in 1932.

Source: Unknown.

Literature:

Burchard, 1884: Placer occurrence in Chloride district.
U.S. Bur. Mines, 1932–33: Lists production for Chloride.

SOCORRO COUNTY

28. ROSEDALE DISTRICT

Location: East flank of the San Mateo Mountains, T. 6 S., Rs. 5 and 6 W.

Topographic maps: Grassy Lookout and Tenmile Hill 7½-minute quadrangles.

Geologic map: Dane and Bachman, 1961, Preliminary geologic map of the
southwestern part of New Mexico, scale 1:380,160.

Access: From Truth or Consequences, 40 miles north on U.S. Highway 85
to junction with State Highway 107. Rosedale is accessible by 8 miles of
dirt road leading west from State Highway 107, 18 miles northwest of
U.S. Highway 85.

Extent: Unknown. Small placer was probably located in gravels near Rosedale mines.

Production history: Placer gold was recovered from stream gravels by pan-
ning in 1904 and 1905. No other information about the deposit is known.

Source: The ore in the Rosedale district was valuable only for its gold content, and consists of free milling gold in quartz veins in shear zones contained in Tertiary rhyolites. Most of the ore is oxidized and, in its higher
grade, is associated with manganese oxides. The placer gold was probably
derived from these oxidized ores.

Literature:

File, 1965: Reports Burris placer claim in district.

Lasky, 1932: Describes lode deposits and mining history in district.

TAOS COUNTY
29. RIO GRANDE AREA

Location: Valley of the Rio Grande from Embudo north to Red River and Cabresto Creek.

Topographic map: Taos and vicinity 30-minute quadrangle.

Geologic map: Bachman and Dane, 1962, Preliminary geologic map of the northeastern part of New Mexico, scale 1:380,160.

Access: From Taos, different points along the Rio Grande are accessible by local roads from State Highway 3 (north of Taos) and from U.S. Highway 64 (south of Taos).

Extent: The placer deposits occur in the riverbed, flood plain, and terrace gravels along the Rio Grande, from the county line north to the mouth of the Red River, a distance of about 33 miles. Placers also occur along the valleys of Red River, Lama Canyon, Alamo Canyon, Garrapata Canyon, San Cristobal, and the Rio Hondo.

Production history: The placers along the Rio Grande were worked during Spanish colonial times (ca. 1600). Recorded production during this century is negligible, but it is probable that much of the gold recovered was not reported to the U.S. Bureau of Mines. The placers have been the subject of study by many writers and the subject of much speculation and perhaps unwarranted investment in development. Two dredging operations (1902, and 1930's) failed because of the presence of large basalt boulders in the river gravels. The river bar deposits have yielded practically all the meager gold recovered.

Source: Unknown, but probably from gold-bearing veins in the Taos Range, east of the river. Some reports suggest that the gold was derived from Precambrian quartzites.

Literature:

Anderson, 1957: Extent; geology; placer operations.

Carruth, 1910; Summarizes earlier studies; extent and value of placers.

Deane, 1896: Placer-mining operations.

Graton and Lindgren, 1910: Notes placer occurrence.

Howard, 1967: Placer operations; source.

Schilling, 1960: Extent and character of placers; size of gold; mining operations.

Silliman, 1880: Detailed account of extent and nature of these placer gravels with estimates of grade and suggestions for operation.

30. RED RIVER DISTRICT

Location: Area around the town of Red River in the Sangre de Cristo Range, Tps. 28 and 29 N., Rs. 14 and 15 E.

Topographic maps: Taos and vicinity 30-minute quadrangle; Comanche Point and Red River 7½-minute quadrangles.

Geologic maps:

Bachman and Dane, 1962, Preliminary geologic map of the northeastern part of New Mexico, scale 1:380,160.

Schilling, 1960, Geology, mines, and prospects of the Red River mining subdistrict (pl. 1), scale 1¾ in.=1 mile.

Access: From Taos, 25 miles north on State Highway 3 to Questa, and 20 miles east on State Highway 38 to Red River. Roads lead north and south of town.

Extent: Small placers have been worked for many years in streams draining the area around the town of Red River. Near Anchor, 6 miles northeast of Red River, placers have been worked in Bitter Creek and in a side canyon extending northeast from Bitter Creek (approx. secs. 16 and 17, T. 29 N., R. 15 E., projected, Comanche Point quadrangle). Near La Belle (2 miles southeast of Anchor), small placers have been worked intermittently in Gold and Spring Creeks, tributaries to Comanche Creek. Placers have been worked intermittently for many years along Placer Creek south of Red River (T. 28 N., R. 14 E., Red River quadrangle). Gold may occur in Tertiary gravels exposed south of Placer Creek and west of the Red River on Gold Hill (approx. sec. 28, T. 28 N., R. 14 E.).

Production history: Recorded production of placer gold has been small in this century. The placers in Bitter Creek were worked during 1898 and again in the 1940's but production is unknown.

Source: Gold veins are found in both the Precambrian granites and in the Tertiary (Miocene?) quartz monzonites in the Taos Range. Although some placer gold was probably derived from the Precambrian veins, it is thought that most placer gold was derived from the Miocene quartz-pyrite veins.

Literature:

Ellis, 1931: Character of placer gold.

Engineering and Mining Journal, 1899: Placer-mining operations.

Graton and Lindgren, 1910: History; location of placer mining.

Mining Reporter, 1898a: Placer-mining operations.

———— 1898b: Placer-mining operations.

Park and McKinley, 1948: Extent; geology; placer-mining operations.

Schilling, 1960: Location; history; type of placer gravels; placer-mining operations; source.

31. RIO HONDO DISTRICT

Location: At the foot of the Sangre de Cristo Range between Arroyo Hondo and Lucero Creek, Tps. 26 and 27 N., R. 13 E. (projected).

Topographic maps: Arroyo Seco and Taos 7½-minute quadrangles.

Geologic maps:

Bachman and Dane, 1962, Preliminary geologic map of the northeastern part of New Mexico, scale 1 : 380,160.

Schilling, 1960, Geology, mines and prospects of the Rio Hondo mining district (pl. 2), scale 1⅛ in.=1 mile.

Access: From Taos, light-duty roads lead north about 8 miles to Arroyo Seco (town) between Rio Lucero and Rio Hondo. Dirt roads lead up these rivers.

Extent: Little is known about the exact occurrence and extent of these placers; apparently they were small and low grade. The placer area was in shallow surface gravels of small debris fans between Lucero Creek and Arroyo Hondo.

Production history: No recorded production.

Source: Unknown; see Red River district

Literature:

File and Northrup, 1966: States that placer gold was found in 1826.

Graton and Lindgren, 1910: Notes placer occurrence.

Howard, 1967: History.

Schilling, 1960: Gold values per cubic yard.

32. PICURIS DISTRICT

Location: Southeast side of the Rio Grande, west of the Picuris Mountains.

Topographic map: Carson 7½-minute quadrangle.

Geologic map: Montgomery, 1953, Geologic map of the Picuris Range (pl. 1.), scale 1:48,000.

Access: Placers are probably located in the vicinity of Pilar at the junction of State Highway 96 and U.S. Highway 64, 16 miles south of Taos.

Extent: Unkown. Placer gold recovered by the La Grande Gold Mining Co. was credited to the Picuris district; the probable location of the placer is along the Rio Grande in the vicinity of Pilar (sec. 32, T. 24 N., R. 11 E.), northwest of the Picuris Mountains.

Production History: The La Grande Gold Mining Co, produced gold from stream gravels worked by sluicing in 1908. The same company apparently installed a dredge to work gravels in the same area, or at Tres Piedras—26 miles northwest in the Tusas Mountains—in 1907. (Early mining records are frequently imprecise about locations.) Three miles southwest of Pilar, at a locality called Glenwoody, the "Oro Grande" of Pennsylvania (this company and the "La Grande" Co. are almost certainly the same) had made plans in 1902 to construct a dredge to work the river gravels above Glenwoody.

Source: Unknown. Small scattered quartz veins containing gold occur in the Precambrian rocks of the Picuris Mountains and could have supplied the gold recovered from the placers. At Glenwoody, gold reportedly occurs in a quartzite exposed in the Rio Grande cliff, and the Glen-Woody

Mining and Milling Co. was formed in 1902 to mine this supposedly large low-grade deposit. Estimates of gold values as high as $1.40 to $3 per ton were made, but mill returns apparently amounted to only 40¢ per ton and lower.

Literature:

Graton and Lindgren, 1910: Mining history at Glenwoody camp; gold values in quartzite.

Howard, 1967: History of placer mining at Glenwoody (under Rio Grande Placer Region).

Schilling, 1960: History of placer mining at Glenwoody.

U.S. Geological Survey, 1907, 1908: Placer-mining activity; operations of La Grande Gold Mining Co.

UNION COUNTY

33. FOLSOM AREA

Location: Cimarron River Valley, northeast of Folsom, T. 31 N., R. 31 E.

Topographic map: Dalhart 2-degree sheet, Army Map Service.

Geologic map: Bachman and Dane, 1962, Preliminary geologic map of the northeastern part of New Mexico, scale 1:380,160.

Access: State Highway 325 leads 20 miles northeast from Folsom to the Cimarron River Valley and placer area.

Extent: Gravels derived from erosion of basalt lava in the Cimarron River Valley contain small amounts of placer gold. The basalt is about 20 miles northeast of Folsom and flowed within the river valley for several miles. The exact location of the gold veinlets and placers near the flow is not known.

Production history: Apparently, some gold was recovered from the gravels as it is described as small flattened grains, but no production has been recorded from the area. The placer is not commercial.

Source: Small gold veinlets in basalt.

Literature:

Anderson, 1957: Notes reports of placer gold.

Harley, 1940: Describes placer occurrences.

OTHER PLACER DEPOSITS

Various surveys of New Mexico mineral resources mention the occurrence of placers in many districts or areas not described in the present report, and frequently these citations are repeated in later publications. I have consulted the literature describing those areas where placers have been reported, but have found no description of any placers. There may be small concentrations of gold in gravel deposits near many lode mines which were worked by miners for a short time, but these occurrences probably were so minor that they received only passing attention.

Chaves County.—Schrader, Stone, and Sanford (1916, p. 214), report gold along the Rio Hondo, presumably derived from deposits in Lincoln County to the west.

Grant County.—Anderson (1957, p. 20) and Lasky and Wooton (1933, p. 39) report a "Gold Camp" in Grant County; no other mention has been found of a Gold Camp in this county. A Gold Camp does occur in the Organ Mountains in Dona Ana County, but placer deposits are not mentioned for this district.

Harding County.—Harley (1940, p. 45) reports minor gold placers in the valley of Ute Creek near Gallegos, apparently discovered in the 1930's. He states that these were probably derived from basaltic flows to the north.

Hildalgo County.—Anderson (1957, p. 20) and Lasky and Wooton (1933, p. 39) list Lordsburg as a placer locality. Lordsburg has received much attention by many writers as a famous mining camp, but no mention has been found of location of placer deposits.

Quay County.—Jones (1904, p. 19) points out that a much-publicized "gold strike" in January 1904 was a hoax. The gravels on Reveulto Creek, 18 miles east of Tucumcari, were salted.

Socorro County.—Howard (1967, p. 169) reports a Kolosa placer claim in T. 5 S., R. 6 E., in the Mound Springs district (west of the Sierra Oscura) midway between Estey City and the Jones district. No information about this placer has been found.

File (1965, p. 71) reports Placer King operations in the Silver Mountain district; the Water Canyon district (T. 3 S., R. 3 W.) is also known as the Silver Mountain district. Lasky (1932, p. 46–54) discusses the Water Canyon district and reports that sparse silver and gold mineralization is present in volcanic rocks; native gold is said to occur in a narrow vein at the Maggie Merchant claim in Shakespeare Gulch, where it is associated with galena, sphalerite, and chalcopyrite.

GOLD PRODUCTION FROM PLACER DEPOSITS

New Mexico rates ninth in the United States (seventh in the western continental States) in placer gold production. The U.S. Bureau of Mines (1967, p. 15) cites 505,000 troy ounces of placer gold produced in New Mexico from 1792–1964. However, I estimate a larger placer gold production. The largest producing districts were Elizabethtown district (Colfax County), Pinos Altos district (Grant County), Old and New Placers districts (Santa Fe County) and Hillsboro district (Sierra County). All sources of information indicate that the Elizabethtown placers have had the largest production, estimated at $5 million (Howard, 1967). During and after the depression years, the Hillsboro district was the largest producer in the State. Table 1 gives the available production information for 33 placer districts. For comparison, I have included, as table 2, a list of 17 gold districts in New Mexico that have produced more than 10,000 ounces of gold (from Koschmann and Bergendahl, 1968).

TABLE 1.—*New Mexico placer gold production, in ounces*

Map locality (pl. 1)	County and placer district	Estimated production, discovery to 1902	Recorded production (data from U.S. Bur. Mines)			Total recorded production 1902–68	Total estimated production	Reference source for estimated production
			1902–33	1934–42	1943–68			
	Colfax:							
1, 2	Elizabethtown and Mount Baldy.	225,000	23,508	1,611	48	25,167	250,000	Howard (1967).
3	Cimarroncito	Unknown	0	0	0	0	Unknown	
	Grant:							
4	White Signal	Minor	21	345	0	366	<1,000	Lasky and Wooton (1933).
5	Pinos Altos	38,842	4,122	1,771	102	5,995	50,000	
6	Bayard area	Unknown	30	79	19	128	<1,000	
	Hidalgo:							
7	Sylvanite	0	106	3	0	109	<200	
	Lincoln:							
8	Jicarilla	4,500	1,150	1,868	2	3,020	8,000	Do:
9	White Oaks	Unknown	860	25	0	885	1,000	
10	Nogal	Unknown	50	84	0	134	200	
	Mora:							
11	Mora River placers	0	0	0	0	0	0	
	Otero:							
12	Orogrande	400	972	564	10	1,546	>2,000	Lindgren, Graton, and Gordon (1910).
	Rio Arriba:							
13	Hopewell	15,000	81	40	0	121	~16,000	Do:
14	El Rito region	Unknown	0	0	0	0	0	
15	Rio Chama	Unknown	>0	0	0	>0	<100	
	Sandoval, Bernalillo, and Valencia:							
16	Placitas-Tejon	Unknown	49	0	0	49	>50	
17	Tijeras Canyon	Unknown	0	0	0	0	>50	

Area							Notes
18 Hell Canyon	Unknown	0	0	0	0	>50	
San Miguel:							Howard (1967).
19 Willow Creek	Unknown	0	0	0	0	100	
20 Villanueva	0	0	0	0	0	0	
21 Las Vegas	0	0	0	0	0	0	
Santa Fe:							
22 Old Placers	100,000	193	1,348	17	1,558	>100,000	Koschmann and Bergandahl (1968):
23 New Placers	96,759	2,117	555	53	2,725	>100,000	
24 Santa Fe	Unknown	0	0	0	0	0	
Sierra:							
25 Hillsboro	104,000	2,016	13,357	186	15,559	120,000	Harley (1934).
26 Pittsburg	0	3,444	3,645	0	7,089	8,000	
27 Chloride	0	2	0	0	2	2	
Socorro:							
28 Rosedale	0	15	0	0	15	15	
Taos:							
29 Rio Grande area	Unknown	7	9	0	16	<1,000	
30 Red River	Unknown	100	5	0	105	<500	
31 Rio Hondo	Unknown	15	0	0	15	<500	
32 Picuris	Unknown	65	0	0	65	65	
Union:							
33 Folsom	Unknown	0	0	0	0	0	
Total	584,501	38,923	25,309	437	64,669	661,000	
Undistributed	-----	906	0	66	972	-----	
State total	584,501	39,829	25,309	503	65,641	661,000	

TABLE 2.—*Major gold districts in New Mexico*

[From Koschmann and Bergendahl (1968). Production, in ounces, to 1959]

County	District	Lode production	Placer production
Bernalillo	Tijeras Canyon	34, 488	
Catron	Mogollon	362, 225	
Colfax	Elizabethtown-Baldy	221, 400	[1] 145, 138
Dona Ana	Organ	11, 435	
Grant	Central	140, 000	
	Pinos Altos	104, 975	[1] 42, 647
	Steeple Rock	>34, 050	
Hidalgo	Lordsburg	223, 750	
Lincoln	White Oaks	<146, 500	[2]
	Nogal	<12, 850	[2]
Otero	Jarilla	<16, 500	[2]
Sandoval	Cochiti	41, 500	
San Miguel	Willow Creek	178, 961	
Santa Fe	Old Placer	[3]	99, 300
	New Placer	16, 000	99, 690
Sierra	Hillsboro	>[3]50, 000	>106, 400
Socorro	Rosedale	27, 750	15

[1] See table 1 for different estimate of placer gold production.
[2] See table 1 for placer gold production; Koschmann and Bergendahl do not give placer production separately:
[3] Koschmann and Bergendahl do not give lode production separately.

Lack of water available for mining purposes has hindered production in most mining districts. Large-scale dredges have operated in only a few districts: Elizabethtown district on the Moreno River, 1901–03; Hillsboro district on Gold Run Creek, 1935–42; Pinos Altos district on Bear Creek and Santo Domingo Creek, 1939–41. Most placer mining in the State was done by individuals working with small equipment such as drywashing jigs.

SUMMARY

The ultimate source of detrital gold in placer deposits is, for the most part, gold-bearing lode deposits of various types, which in New Mexico are represented by fissure veins and disseminated ores of Precambrian and Tertiary ages. The most productive lode deposits throughout the State are the fissure veins in Tertiary intrusive rocks; in some districts, gold-bearing veins in associated contact-metamorphic rocks have also yielded appreciable quantities of gold.

Most placers in New Mexico are erosional products of gold-bearing Tertiary lodes. Lindgren, Graton, and Gordon (1910, p. 75–76) emphasize the importance of the derivation of placer deposits from lode deposits of early Tertiary age. Unfortunately, I have found no information relating to absolute age dates on any of the source rocks in New Mexico. However, geologic studies have indicated that the age of the source rocks for most of the major districts is younger than Lindgren thought and probably extends from Eocene to Miocene time. The intrusion that formed Mount Baldy, Colfax County, cut Upper Cretaceous and Paleocene sedimentary rocks and therefore is younger than Paleocene. Gold-pyrite veins cut Miocene(?) intrusive rocks which have formed minor placer deposits in Taos County, 10 miles west of Mount Baldy. Gold-bearing veins in Oligocene volcanic rocks have been eroded to form rich placer deposits in the Hillsboro district of Sierra County. The intrusive rocks in the San Pedro district, Santa Fe County, are considered to be early Oligocene to early Miocene in age; the similarity between these and other intrusive rocks in the Ortiz Mountains suggests that this porphyry belt is generally the same age.

Most of the placer deposits of the State were formed during the Quaternary in parts of alluvial fans and in drainages leading from adjacent mineralized areas. The gold contained in the gravels is characteristically angular, indicating proximity to the source. However, the placers on the Mora River, and probably the placers in the Rio Grande, bear gold that appears to have traveled a long distance.

Only a few deposits are the product of erosional cycles earlier than Quaternary. Erosion has been continuous since late Tertiary in the Elizabethtown-Baldy Mountain districts (Colfax County) and the Old Placers-New Placers districts (Santa Fe County); it is likely that Tertiary placer deposits were mined in these districts. Some of the placer deposits of the Pittsburg district (Sierra County) are found in the late Tertiary Santa Fe Group; gold is also found in the Cambrian Bliss Sandstone at the Caballo Dam. The placers of the Sandia Mountains may be reworked sediments from the Santa Fe Group. The Tertiary Santa Fe conglomerate at El Rito (Rio Arriba County) and the Permian Glorieta Sandstone at Villanueva (San Miguel County) are auriferous. An auriferous conglomerate of probable Tertiary age at Hopewell (Rio Arriba County) is overlain by gravels of recent age, making it the only known economic buried stream-channel placer in the State.

Although it is true that the most productive gold placer districts of New Mexico are adjacent to the principal gold lode districts, several gold lode districts have no associated placers. A review of the literature of important lode districts suggests that two complementary processes contribute to the deposition of placer gravels.

Oxidation of gold-bearing sulfide ores helps free the gold and facilitates its erosion and sedimentary reconcentration. Districts where free gold is known to occur in the oxidized parts of fissure veins include Pinos Altos, Bayard, White Oaks, Nogal, Jicarilla, Red River, Orogrande, and Hopewell. In the Hopewell district (Rio Arriba County) and the Baldy Mountain district (Colfax County), coarse free gold is seen in quartz-pyrite veins. The ores are found at the surface in most districts, and there are several indications that large parts of many ore bodies were eroded throughout the Quaternary.

The Mogollon district (Catron County) has the largest recorded gold production in New Mexico. No placer production has been recorded from this district. The Lordsburg district (Hidalgo County) and the Willow Creek district (San Miguel County) have been mentioned as placer districts, but no placer gold production has been recorded for either. In addition, the Central district (Grant County) and the White Oaks district (Lincoln County) have produced more than 100,000 ounces of lode gold but only small amounts of placer gold.

Most of the gold produced in the Central district was a byproduct of mining base-metal ores; the small placers in the vicinity of Bayard (p. 10) were the product of erosion of gold-bearing sulfide veins situated in a very small area. No placers have been reported associated with base-metal deposits in the surrounding region. The small placers in the White Oaks district are also found in a very small area; the district, however, is predominantly a gold district, and the deposits are close to a wide alluvial plain. It is possible that gold eroded from the veins was not concentrated but rather was distributed throughout the gravels in this plain.

The gold in the Lordsburg district has been produced primarily as a byproduct of base-metal ores. The Lordsburg district is composed of two subdistricts, the Virginia and the Pyramid. In the Virginia subdistrict secondary alteration and oxidation of the ores is erratic; in the Pyramid subdistrict, Quaternary alluvium covers large parts of the area. Lack of oxidation of the ores would account for the lack of placer deposits in the Virginia subdistrict, and burial by alluvial cover or distribution of eroded gold may account for lack of placer deposits in the Pyramid subdistrict.

New Mexico has been thoroughly prospected for gold deposits since 1828; all extensively exposed placer deposits have probably been found. However, there is indication that not all these deposits have been thoroughly explored and (or) exploited. Areas of placer concentration are apparently unmined at Jicarilla, Old Placers, Orogrande, Hillsboro, Pittsburg, and Hopewell districts. Extensive testing of these unmined gravels might reveal large tonnages of suitable grade to warrant future mining.

BIBLIOGRAPHY

LITERATURE REFERENCES

Anderson, E. C., 1955, Mineral Deposits and mines in south-central New Mexico, *in* Guidebook of south-central New Mexico: New Mexico Geol. Soc., 6th Field Conf., p. 121–122.

———— 1956, Mining in the southern part of the Sangre de Cristo Mountains, *in* Guidebook of southeastern Sangre de Cristo Mountains, New Mexico: New Mexico Geol. Soc. 7th Field Conf., p. 139–142.

———— 1957, The metal resources of New Mexico and their economic features through 1954: New Mexico Bur. Mines and Mineral Resources Bull. 39, 183 p.

General description of mining areas of the State.

Atkinson, W. W., Jr., 1961, Geology of the San Pedro Mountains, Santa Fe County, New Mexico: New Mexico Bur. Mines Bull. 77, 50 p.

Ballman, D. L., 1960, Geology of the Knight Peak area, Grant County, New Mexico: New Mexico Bur. Mines Bull. 70, 39 p.

Bancroft, H. H., 1889, History of Arizona and New Mexico, v. 17 *of* Works of Hubert Home Bancroft: San Francisco, Calif., A. L. Bancroft & Co., 829 p.

Benjovsky, T. D., 1945, Reconnaissance survey of the Headstone (Hopewell) mining district, Rio Arriba County, New Mexico: New Mexico Bur. Mines and Mineral Resources Circ. 11, 10 p.

Mining during Mexican government (p. 340); New Mexico (p. 629–801); early history of the State with valuable information on early mining developments.

Bergendahl, M. H., 1965, Gold, *in* Mineral and water resources of New Mexico: New Mexico Bur. Mines and Mineral Resources Bull. 87, 437 p.

Brief summary of placer mining in the State.

Bingler, E. C., 1968, Geology and mineral resources of Rio Arriba County, New Mexico: New Mexico Bur. Mines and Mineral Resources Bull. 91, 158 p.

Blake, W. P., 1859, Observations of the mineral resources of the Rocky Mountain chain, near Santa Fe, and the probable extent southward of the Rocky Mountain gold field: Boston Soc. Nat. History Proc. v. 7, p. 64–70.

Brinsmade, R. B. 1908, Development of San Pedro Mountain, New Mexico: Mining World, v. 28, p. 1021–1024.

Burchard, H. C., 1882, Report of the Director of the Mint upon the statistics of the production of the precious metals in the United States (for the year 1881): Washington, 765 p. [New Mexico, p. 327–353].

———— 1883, Report of the Director of the Mint upon the statistics of the production of the precious metals in the United States (for the year 1882): Washington, 873 p. [New Mexico, p. 339–389].

———— 1884, Report of the Director of the Mint upon the production of the precious metals in the United States during the calendar year 1883: Washington, 858 p. [New Mexico, p. 562–610].

———— 1885, Report of the Director of the Mint upon the production of the precious metals in the United States during the calendar year 1884: Washington, 644 p. [New Mexico, p. 373–395]

Information and statistics related to placer mining in different districts for each year.

Bush, F. V., 1915, Mining in the Pinos Altos district of New Mexico: Mining World, v. 42, p. 165–168.

Carruth, J. A., 1910, New Mexico gold gravels [Rio Grande placers]: Mines and Minerals, v. 31, p. 117–119.

Chase, C. A., and Muir, Douglas, 1923, The Aztec mine, Baldy, New Mexico [abs.]: Mining and Metallurgy no. 190, p. 33–35.

Deane, C. A., 1896, Placer deposits in New Mexico [Rio Grande Placers]: Mining Industry and Review, v. 16 [Feb. 13, 1896], p. 371–372.

Dinsmore, C. A., 1908, The new gold camp of Sylvanite, New Mexico: Mining World, v. 29, p. 670–671.

Ellis, R. W., 1931, The Red River lode of the Moreno Glacier: New Mexico Univ. Bull., v. 4, no. 3, 26 p.

Elston, W. E., 1967, Summary of the mineral resources of Bernalillo, Sandoval and Santa Fe Counties: New Mexico Bur. Mines Bull. 81, 81 p.

Endlich, F. M., 1883, The mining regions of southern New Mexico [Hillsboro district]: Am. Naturalist, v. 17, p. 149–157.

Engineering and Mining Journal, 1899, Developments in northern New Mexico: Eng. and Mining Jour., v. 68, pt. 1, p. 393.

File, L. A., 1965, Directory of mines in New Mexico: New Mexico Bur. Mines and Mineral Resources Circ. 77, 188 p.

   Lists mines and placer deposits alphabetically (p. 1–97). Locates them by mining district and county and refers briefly to other sources.

File, Lucien, and Northrop, S. A., 1966, County, Township and Range locations of New Mexico's mining districts: New Mexico Bur. Mines and Mineral Resources Circ. 84, 66 p.

   Includes list of district names, synonyms, and older names.

Frost, Max, 1905, Mining in New Mexico [Elizabethtown district]: Mining World, v. 23, p. 6–9.

Gardner, E. D., and Allsman, P. T., 1938, Power shovel and dragline placer mining: U.S. Bur. Mines Inf. Circ. 7013, 68 p.

Gifford, A. W., 1899, Wonderful dry placers [Orogrande district]: Ores and Metals, v. 8, Oct., p. 15.

Gillerman, Elliot, 1964, Mineral deposits of Western Grant County, New Mexico: New Mexico Bur. Mines and Mineral Resources Bull. 83, 213 p.

Gordon, C. H., 1910, Sierra and central Socorro Counties, in Lindgren, Waldemar, Graton, L. C., and Gordon, C. H., The ore deposits of New Mexico: U.S. Geol. Survey Prof. Paper 68, p. 213–285.

Graton, L. C., 1910, Colfax County, in Lindgren, Waldemar, Graton, L. C., and Gordon, C. H., The ore deposits of New Mexico: U.S. Geol. Survey Prof. Paper 68, p. 91–108.

Graton, L. C., 1910a, Lincoln County, in Lindgren, Waldemar, Graton, L. C., and Gordon, C. H., The ore deposits of New Mexico: U.S. Geol. Survey Prof. Paper 68, p. 175–184.

——— 1910b, Otero County, in Lindgren, Waldemar, Graton, L. C., and Gordon, C. H., The ore deposits of New Mexico: U.S. Geol. Survey Prof. Paper 68, p. 184–190.

——— 1910c, Rio Arriba County, in Lindgren, Waldemar, Graton, L. C., and Gordon, C. H., The ore deposits of New Mexico: U.S. Geol. Survey Prof. Paper 68, p. 124–133.

Graton, L. C., and Lindgren, W., 1910, Taos County, in Lindgren, Waldemar, Graton, L. C., and Gordon, C. H., The ore deposits of New Mexico: U.S. Geol. Survey Prof. Paper 68, p. 82–91.

Graton, L. C., Lindgren, Waldemar, and Hill, J. M., 1910, Grant County, in Lindgren, Waldemar, Graton, L. C., and Gordon, C. H., The ore deposits of New Mexico: U.S. Geol. Survey Prof. Paper 68, p. 295–348.

Griswold, G. B., 1959, Mineral deposits of Lincoln County, New Mexico: New Mexico Bur. Mines Bull. 67, 117 p.

Harley, G. T., 1934, Geology and ore deposits of Sierra County, New Mexico: New Mexico Bur. Mines Bull. 10, 220 p.

——— 1940, The geology and ore deposits of northeastern New Mexico (exclusive of Colfax County): New Mexico Bur. Mines and Mineral Resources Bull. 15, 102 p.

Harrington, E. R., 1939, Gold mining in the desert [New Placer and Old Placer district]: Mines Mag., v. 29, p. 508–509, 512.

Hartly, Carney, 1915, The opportunity in placer mining [Old Placer district]: Eng. and Mining Jour., v. 99, p. 185–188.

Heikes, V. C., and York, C. G., 1913, Dry placers in Arizona, Nevada, New Mexico, and California: U.S. Geol. Survey Mineral Resources U.S. [1912], pt. 1, p. 254–263.

Hernon, R. M., 1953, Summary of smaller mining districts in the Silver City region, in Guidebook of southwestern New Mexico: New Mexico Geol. Soc., 45th Field Conf. [Oct. 15–18], p. 138–141.

Herrick, C. L., 1898, The geology of the San Pedro and Albuquerque districts: New Mexico Univ. Bull., v. 1, no. 4; New Mexico Univ. Sci. Lab. Div. 7, art. 5, p. 93–116.

Hill, J. M., 1910, Sylvanite district, in Lindgren, Waldemar, Graton, L. C., and Gordon, C. H., Ore deposits of New Mexico: U.S. Geol. Survey Prof. Paper 68, p. 338–343.

Howard, E. V., 1967, Metalliferous occurrences in New Mexico: Santa Fe, N.M., State Planning Office, 270 p.
    Describes mining district of the State alphabetically.

James, Henry, 1955, Muleshoe Gold [New Placer district]: New Mexico Sun Trails, v. 8, no. 1, p. 20–21.

Jones, F. A., 1903, Report of the Director of the Mint upon the production of the precious metals in the United States during the Calendar year 1902: Washington, 391 p. (G. E. Roberts, Director of the Mint) ; [New Mexico, p. 168–177].

——— 1904, New Mexico mines and minerals: Santa Fe, N. Mex., New Mexican Printing Co., 349 p.
    Comprehensive descriptions of mining areas including placers active at that time; includes much information on history of areas.

——— 1906, Placers of Santa Fe County, New Mexico: Mining World, v. 25, p. 425.

——— 1908a, Sylvanite, New Mexico, the new gold camp: Eng. and Mining Jour., v. 86, p. 1101–1103.

——— 1908b, The new camp of Sylvanite, New Mexico: Mining Sci., v. 58, p. 489–491.

——— 1908, Epitome of the economic geology of New Mexico: Albuquerque, New Mexico Bur. Immigration, 47 p.
    An outline of important mineral deposits of the State. Chief placer-mining districts are listed on pages 12–13.

Jones, F. A., 1915, The mineral resources of New Mexico: New Mexico School of Mines, Mineral Resources Survey Bull. 1, 77 p.
    Revised edition of 1908 publication. Placer districts are listed on page 22; map of chief mining districts is included.

Just, Evan, 1937, Geology and economic features of the pegmatites of Taos and Rio Arriba Counties, New Mexico: New Mexico Bur. Mines Bull. 14, 73 p.

Kelley, V. C., 1951, Oolitic iron deposits of New Mexico [Pittsburgh district]: Am. Assoc. Petroleum Geologists Bull., v. 35, p. 2199–2228.

Kelley, V. C., and Silver, Caswell, 1952, Geology of the Caballo Mountains with special reference to regional stratigraphy and structure, and to mineral resources including oil and gas: New Mexico Univ. Pubs. Geology, no. 4, 286 p.

Keyes, C. R., 1903, Geology of the Apache Canyon placers: Eng. and Mining Jour., v. 76, p. 966–967.

Koschmann, A. H., and Bergendahl, M. H., 1968, Principal gold producing districts of the United States: U.S. Geol. Survey Prof. Paper 610, p. 200–211.

Describes seventeen districts in New Mexico which produced more than 10,000 ounces of gold through 1957. Major placer districts are described.

Lasky, S. G., 1932, The ore deposits of Socorro County, New Mexico: New Mexico Bur. Mines and Mineral Resources Bull. 8, 139 p.

Lasky, S. G., 1936, Geology and ore deposits of the Bayard area, Central mining district, New Mexico: U.S. Geol. Survey Bull. 870, 144 p.

———— 1947, Geology and ore deposits of the Little Hatchet Mountains, Hidalgo and Grant Counties, New Mexico: U.S. Geol. Survey Prof. Paper 208, 101 p.

Lasky, S. G., and Wooton, T. P., 1933, The metal resources of New Mexico and their economic features: New Mexico School of Mines Bull. 7, 178 p.

General description of mining areas in the State. Predecessor to Anderson, 1957.

Leatherbee, Brigham, 1911, Mesa del Oro placer grounds [Hillsboro district]: Mining World, v. 35, p. 536.

Lee, W. T., 1916, The Aztec gold mine, Baldy, New Mexico: U.S. Geol. Survey Bull. 620, p. 325–330.

Lindgren, Waldemar, 1910, Santa Fe County, in Lindgren, Waldemar, Graton, L. C., and Gordon, C. H., Ore deposits of New Mexico: U.S. Geol. Survey Prof. Paper 68, p. 163–175.

Lindgren, Waldemar, Graton, L. C., and Gordon, C. H., 1910, The ore deposits of New Mexico: U.S. Geol. Survey Prof. Paper 68, 361 p.

Detailed discussion of geology and lode and placer mining for each county in New Mexico. [Descriptions of individual counties are written by different authors and annotated separately for each county discussed in this paper.]

Metzger, O. H., 1938, Gold mining in New Mexico: U. S. Bur. Mines Inf. Circ. 6987, 71 p.

Placer deposits and operations are discussed for major placer areas.

Mining Reporter, 1898a, New Mexico placer mining: Mining Reporter, v. 38, no. 1, p. 22–23.

———— 1898b, Bitter Creek placers [Red River district]: Mining Reporter, v. 38, no. 6, p. 22.

Northrop, S. A., 1942, Minerals of New Mexico: New Mexico Univ. Bull. 379, Geology Ser., v. 6, no. 1, 387 p.

Partial list of placer gold localities is given on pages 158–159.

Paige, Sidney, 1911, The ore deposits near Pinos Altos, New Mexico: U.S. Geol. Survey Bull. 470, p. 109–125.

———— 1916, Silver City Folio: U.S. Geol. Survey Folio 199, 19 p.

Park, C. F., Jr., and McKinley, P. F., 1948, Geology and ore deposits of Red River and Twinning districts, Taos County, New Mexico: New Mexico Bur. Mines and Mineral Resources Circ. 18, 35 p.

Pettit, R. F., Jr., 1966a, Maxwell Land Grant, in Guidebook to Taos—Raton— Spanish Peaks Country: New Mexico Geol. Soc. 17th Field Conf., p. 66–68.

———— 1966b, History of mining in Colfax County, in Guidebook to Taos—Raton— Spanish Peaks Country: New Mexico Geol. Soc. 17th Field Conf., p. 69–75.

[These two articles in "Guidebook to Taos—Raton—Spanish Peaks Country" are summarized from an open file report by R. F. Pettit, Jr., entitled "Mineral Resources of Colfax County, New Mexico." The present status of the report is that it is on Open File, State Bureau of Mines and Mineral Resources, Socorro, N. Mex.]

Prince, L. B., 1883, Historical sketches of New Mexico: New York, Leggat Bros.; Kansas City, Ramsey, Millett, and Hudson, 327 p.

History of mines and mining given on pages 241–245. Details of history up to 1847.

Ray, L. L., and Smith, J. F., Jr., 1941, Geology of the Moreno Valley, New Mexico: Geol. Soc. America Bull., v. 52, no. 2, p. 177–210.

Raymond, R. W., 1870, Statistics of mines and mining of the States and Territories west of the Rocky Mountains: Washington, 805 p. [New Mexico, p. 381–418].

———— 1872, Statistics of mines and mining in the States and Territories west of the Rocky Mountains for the year 1870: Washington, 566 p. [New Mexico, p. 282–286].

———— 1873a, Statistics of mines and mining in the States and Territories west of the Rocky Mountains (for the year 1871): Washington, 566 p. [New Mexico, p. 337–339].

———— 1873b, Statistics of mines and mining in the States and Territories west of the Rocky Mountains for the year 1872. Washington, 550 p. [New Mexico, p. 309–311].

———— 1874, Statistics of mines and mining in the States and Territories west of the Rocky Mountains for the year 1873: Washington, 585 p. (New Mexico, p. 313–344).

———— 1877, Statistics of mines and mining in the States and the Territories west of the Rocky Mountains for the year 1875: Washington, 519 p. [New Mexico, p. 337–340].
Information and statistics related to placer mining in different districts for each year.

Reiche, Parry, 1949, Geology of the Manzanita and North Manzano Mountains, New Mexico [Hell Canyon district]: Geol. Soc. America Bull., v. 60, no. 7, p. 1183–1212.

Robinson, G. D., Wanek, A. A., Hays, W. H., and McCallum, M. E., 1964, Philmont Country—the rock and landscape of a famous New Mexico ranch [Mount Baldy district]: U.S. Geol. Survey Prof. Paper 505, 152 p.

Schilling, J. H., 1959, Silver City, Santa Rita, Hurley: New Mexico Bur. Mines and Mineral Resources Scenic Trip 5, 43 p.

———— 1960, Mineral resources of Taos County, New Mexico: New Mexico Bur. Mines and Mining Resources Bull. 71, 124 p.

Schmidt, P. G., and Craddock, Campbell, 1964, The geology of the Jarilla Mountains, Otero County, New Mexico: New Mexico Bur. Mines Bull. 82, 55 p.

Schrader, F. C., Stone, R. W., and Sanford, S., 1916, Useful minerals of the United States: U.S. Geol. Survey Bull. 624, 412 p.
Gold placer districts are listed on page 214; many of those named are not described nor mentioned by later authors.

Silliman, Benjamin, Jr., 1880, Report on the newly discovered auriferous gravels of the upper Rio Grande del Norte in the counties of Taos and Rio Arriba, New Mexico: Omaha, Nebr., Herald Pub. House, 34 p.

Smith, E. P., and Dominian, Leon, 1904, Notes on a trip to White Oaks, New Mexico: Eng. and Mining Jour., v. 77, p. 799–800.

Smith, T. E., 1940, A mobile dry placer plant [Old Placers district]: Eng. and Mining Jour., v. 141, no. 6, p. 40.

Statz, B. A., 1912, The New Placer mining district, New Mexico: Mining Sci., v. 66, p. 167.

Stone, G. H., 1899, Dry gold placers of the arid regions: Mines and Minerals, v. 19, p. 397–399.

U.S. Bureau of Mines, 1925–34, Mineral resources of the United States [annual volumes, 1924–31]: Washington, U.S. Govt. Printing Office.

———— 1933–68, Minerals Yearbook [annual volumes, 1932–67]: Washington, U.S. Govt. Printing Office.
Information relating to placers cited in text is referenced by year of pertinent volume.

———— 1967, Production potential of known gold deposits in the United States: U.S. Bur. Mines Inf. Circ. 8331, 24 p.
Lists estimates of total placer gold production in troy ounces.

U.S. Geological Survey, 1896–1900, Annual reports [17th through 21st, 1895–1900]: Washington, U.S. Govt. Printing Office.

———— 1883–1924, Mineral resources of the United States [annual volumes, 1882–1923]: Washington, U.S. Govt. Printing Office.
Information relating to placers cited in text is referenced by year of pertinent volume.

Wells, E. H., 1930, An Outline of the mineral resources of New Mexico: New Mexico Bur. Mines and Mineral Resources Circ. 1, 15 p.

Wells, E. H., and Wooton, T. P., 1932, Gold mining and gold deposits in New Mexico: New Mexico Bur. Mines and Mineral Resources Circ. 5, 24 p. [Revised by T. P. Wooton, April 1940].
Report contains general information on mining and general features of placer deposits. Individual placers are described in detail for most important placer districts. Bibliography of papers relating to lode and placer gold deposits.

Wolle, M. S., 1957, Pinos Altos, New Mexico Gold Camp: Mining World, v. 19, no. 12, p. 56–57.

Wright, I. L., 1915, The Pinos Altos district, New Mexico: Eng. and Mining Jour., v. 99, p. 133–135.

Wright, P. E., 1932, The Jicarilla mining district of New Mexico: Mining Jour. [Phoenix, Ariz.], v. 16, no. 8, p. 7.

Yung, M. B., and McCaffery, R. S., 1903, The ore deposits of the San Pedro district, New Mexico: Am. Inst. Mining Engineers Trans., v. 33, p. 350–362.

GEOLOGIC MAP REFERENCE

[References keyed by number to districts given in text]

Atkinson, W. W., Jr., 1961, Geology of the San Pedro Mountains, Santa Fe County, New Mexico: New Mexico Bur. Mines Bull. 77, 50 p., pl. 4.
No. 23.

Bachman, G. O., and Dane, C. H., 1962, Preliminary geologic map of the northeastern part of New Mexico: U.S. Geol. Survey Misc. Geol. Inv. Map I–358, scale 1:380,160.
Nos. 1, 2, 3, 11, 19, 21, 24, 29–31, 33.

Ballman, D. L., 1960, Geology of the Knight Peak area, Grant County, New Mexico: New Mexico Bur. Mines Bull. 70, 39 p., pl. 1.
No. 4.

Bingler, E. C., 1968a, Geology and mineral resources of Rio Arriba County, New Mexico: New Mexico Bur. Mines and Mineral Resources Bull. 91, 158 p., pls. 1, 4.
Nos. 13, 15.

———— 1968b, Reconnaissance geology of the El Rito 7½-minute quadrangle: New Mexico Bur. Mines and Mineral Resources Geol. Map 20.
No. 14.

———— 1968c, Reconnaissance geology of the Valle Grande Peak 7½-minute quadrangle: New Mexico Bur. Mines and Mineral Resources Geol. Map 21.
No. 14.

Dane, C. H., and Bachman, G. O., 1957, Preliminary geologic map of the northwestern part of New Mexico: U.S. Geol. Survey Misc. Geol. Inv. Map. I–224, scale 1:380,160.
Nos. 16–18, 22, 23.

———— 1958, Preliminary geologic map of the southeastern part of New Mexico: U.S. Geol. Survey Misc. Geol. Inv. Map I–256, scale 1:380,160.
Nos. 8–10.

———— 1961, Preliminary geologic map of the southwestern part of New Mexico: U.S. Geol. Survey Misc. Geol. Inv. Map I–344, scale 1:380,160.
Nos. 4, 5, 6, 7, 25–28.

Gillerman, Elliot, 1964, Mineral deposits of Western Grant County, New Mexico: New Mexico Bur. Mines and Mineral Resources Bull. 83, 213 p., pl. 1.
No. 4.

Griswold, G. B., 1959, Mineral deposits of Lincoln County, New Mexico: New Mexico Bur. Mines Bull. 67, 117 p., pl. 2; figs. 21, 2.
Nos. 8–10.

Harley, G. T., 1934, Geology and ore deposits of Sierra County, New Mexico: New Mexico Bur. Mines Bull. 10, 220 p., pls. 6, 7.
Nos. 25, 26.

Johnson, R. B., 1970, Geologic map of the Villanueva quadrangle, San Miguel County, New Mexico: U.S. Geol. Survey Geol. Quad. Map GQ–869, scale 1:62,500.
No. 20.

Kelley, V. C., and Silver, Caswell, 1952, Geology of the Caballo Mountains with special reference to regional stratigraphy and structure, and to mineral resources including oil and gas: New Mexico Univ. Pubs. Geology, no. 4, 286 p., fig. 2.
No. 26.

Kuellmer, F. J., 1956, Geologic map of Hillsboro Peak: New Mexico Inst. Mining and Technology, Geologic Map 1, 30-minute quad.
No. 25.

Lasky, S. G., 1936, Geology and ore deposits of the Bayard area, Central mining district, New Mexico: U.S. Geol. Survey Bull. 870, 144 p., pls. 1, 9.
No. 6.

———— 1947, Geology and ore deposits of the Little Hatchet Mountains, Hidalgo and Grant Counties, New Mexico: U.S. Geol. Survey Prof. Paper 208, 101 p., pl. 1; fig. 2.
No. 7.

Miller, J. P., Montgomery, Arthur, and Sutherland, P. K., 1963, Geology of part of the southern Sangre de Cristo Mountains, New Mexico: New Mexico Bur. Mines and Mineral Resources Mem. 11, 106 p., pl. 1.
No. 19.

Montgomery, Arthur, 1953, Pre-Cambrian geology of the Picuris Range, north-central New Mexico: New Mexico Bur. Mines and Mineral Resources Bull. 30, 89 p.
No. 32.

Paige, Sidney, 1911, The ore deposits near Pinos Altos, New Mexico: U.S. Geol. Survey Bull. 470, p. 109–125, fig. 10.
No. 5.

Ray, L. L., and Smith, J. F., Jr., 1941, Geology of the Moreno Valley, New Mexico: Geol. Soc. America Bull., v. 52, no. 2, p. 177–210, pls. 1, 2.
No. 1.

Reiche, Parry, 1949, Geology of the Manzanita and North Manzano Mountains, New Mexico: Geol. Soc. America Bull., v. 60, no. 7, p. 1183–1212, pl. 5.
No. 18.

Schilling, J. H., 1960, Mineral resources of Taos County, New Mexico: New Mexico Bur. Mines and Mining Resources Bull. 71, 124 p., pl. 1.
Nos. 30, 31, 32.

Schmidt, P. G., and Graddock, Campbell, 1964, The geology of the Jarilla Mountains, Otero County, New Mexico: New Mexico Bur. Mines Bull. 82, 55 p., pl. 1.
No. 12.

Smith, C. T., and Budding, A. J., 1959, Little Black Peak, east half: New Mexico Inst. Mining and Technology, Geol. Map 11.
No. 9.

Wanek, A. A., Read, C. B., Robinson, G. D., Hays, W. H., and McCallum, Malcolm, 1964, Geologic map and sections of the Philmont Ranch region, New Mexico: U.S. Geol. Survey Misc. Geol. Inv. Map I–425, scale 1:48,000 (see also Robinson and others, 1964, pls. 3, 5).
Nos. 2, 3.